D0554161

)))

RADIO
AND TELEVISION
REGULATION

Radio and Television Regulation

Broadcast Technology in the United States, 1920–1960

Hugh R. Slotten

The Johns Hopkins University Press
Baltimore and London

Printed in the United States of America on acid-free paper
9 8 7 6 5 4 3 2 1

The Johns Hopkins University Press
2715 North Charles Street
Baltimore, Maryland 21218-4363
www.press.jhu.edu

Library of Congress Cataloging-in-Publication Data will be found at the end of this book.
A catalog record for this book is available from the British Library.

ISBN 0-8018-6450-X

)))

CONTENTS

)))

PREFACE AND ACKNOWLEDGMENTS

Before the 1970s and 1980s, few government officials seriously questioned the regulatory policies established as early as the 1920s and 1930s for the broadcast industry in the United States. The 1934 Communications Act stipulated that broadcasters had an obligation to act in the public interest as public trustees of the airwaves. The law directed the federal government to regulate the industry by making sure its decisions benefited the U.S. public. But especially since the 1980s, policies based on the concept of deregulation have become prevalent. For example, the federal government has relaxed regulations limiting station ownership by single companies, and broadcasters no longer feel pressured to survey community needs for "public interest" programming. The 1996 Telecommunications Act has given formal statutory authority to many of these new decisions. Industry and government officials argue that new policies are necessary to prepare the country for new technological developments connected with the establishment of a "national information infrastructure" or "national information superhighway." In order to facilitate the predicted convergence of television, telephone, and computer technology, the new law stresses the importance of marketplace competition instead of active government regulation.

Despite these new developments, earlier policies and traditional themes remain highly relevant. Decisions about national policy for broadcasting still depend on the resolution of competing understandings of the public interest as defined by government officials, industry representatives, and interested citizens. Fundamental concerns from earlier periods about such issues as economic concentration and monopoly control of information still exist. Although the federal government has decided for the first time to treat parts of the electromagnetic spectrum as a private property resource by allowing cellular-phone companies and other private businesses to submit bids for the use of channels, officials have resisted calls to force broadcasters to pay for the use of their frequencies. Broadcasters are still seen as public trustees of the airwaves. If they do not operate in the public interest, at least theoretically, their spectrum rights can be withdrawn. The federal government also continues to play an important role in the development and implementation of new technologies. This is particularly evident in the recent decisions mandating the conversion of television to an interactive high-definition/digital standard consistent with the computer technology necessary for a national information infrastructure.

Only by gaining a deeper understanding of the early role of the federal government in the development of the broadcast industry and of the early traditions established can we gain a critical comprehension of current problems. A historical analysis of broadcast policy is especially important because current debates about deregulation tend to assume a sharp dichotomy between private and public sectors. They often ignore the associational and cooperative activities between government and industry that were crucial to the development of broadcasting in the United States. This book examines key decisions made by the government institutions that oversaw the development of the radio and television industry, including the Department of Commerce from 1910 to 1927, the Federal Radio Commission (FRC) from 1927 to 1934, and the Federal Communications Commission (FCC) after 1934. It covers the principal decades during which regulation was the guiding philosophy, primarily before the 1970s, when officials first questioned these policies. The work specifically focuses on the intersection of technical issues and the social, political, legal, and economic components of decision making by policy makers of the four most important broadcast technologies developed during the first half of the twentieth century: AM (amplitude modulation) radio and FM (frequency modulation) radio, and monochrome television and color television. The book analyzes the policy debates during the different periods when the public implications of each new technology first became important, examining specifically the role of the different participants and the significance of key policy decisions for the early commercial development of the technologies.

The broad approach presented here attempts to bridge an important gap between studies in the history of technology and science that do not adequately engage the role of the state as well as other policy-making institutions and policy studies, historical or otherwise, that fail to deal with standardization and other technical developments adequately. The focus on engineering standards is particularly important for illuminating key developments relating to the use of experts in integrating liberal ideals within a corporate economy dependent on consumer demand. In addition to its broad historiographic significance (as well as its relevance for current policy debates), this book aims to make a major contribution to the literature on the history of broadcasting and the broadcast industry. No other work has made a similar extensive use of archival sources for this entire historical period.[1]

The first chapter provides an overview of the development of national policy for AM radio before 1927. Although amateur radio enthusiasts had made a number of earlier efforts to use radio to broadcast to the public, the first major attempt occurred in 1920, when KDKA in Pittsburgh (owned by the Westinghouse Electric and Manufacturing Company) broadcast the presidential election returns. KDKA's success led to the phenomenal expansion of radio. At the beginning of 1921, 28 licensed stations were operating in the United States; by the end of the year, there were more than 550. The rapid expansion of this new public technology presented a number of challenges. Most important, the different stations competed for prime channel assignments on a limited broadcast spectrum. Nearly all participants agreed that government regulation was necessary to prevent chaos on the airwaves. Secretary of Commerce Herbert Hoover established important precedents during the 1920s, when he used his department to regulate broadcasting. Many of the radio engineers who provided technical advice to the FRC and the FCC also served as consultants to the government during this early period. The chapter introduces the major themes of this study in the context of Hoover's efforts to regulate radio by forging an alliance between private industry and the federal government.

Chapter 2 focuses on the regulation of AM radio by the radio commission during the period from 1927 to 1934. After a court ruled in 1926 that Hoover lacked legal authority, radio stations made their own choices of broadcast frequencies and station power. The result was near chaos for radio broadcasting; in some areas, interference made listening nearly impossible. This event also forced Congress finally to act to prevent complete disorder. The Radio Law of 1927 established the radio commission as an independent agency to regulate the industry. The commission sought to rationalize radio broadcasting by imposing order on a chaotic situation. Stations that could not demonstrate they were serving the public interest faced revocation of their licenses. By helping to reinforce particular trends already underway in the growth of the industry, the radio commission was instrumental in establishing the political, economic, administrative, and social characteristics of the U.S. system of broadcasting. The 1934 Communications Act, which established the Federal Communications Commission, sought to centralize the government's administration of communications in one federal agency; as far as the regulation of radio broadcasting was concerned, the 1934 law essentially institutionalized the policies of the radio commission.

Chapter 3 examines the development of public policy for monochrome television during the late 1930s and early 1940s. The FCC was responsible for deciding crucial issues such as the allocation of spectrum space and the establishment of system standards. The resolution of these issues had important implications for the development of the television industry. Having invested more than $1 million in the development of television, the Radio Corporation of America (RCA), in the mid 1930s, pressured the commission immediately to authorize commercial broadcasting. Other manufacturers lobbied the commission to postpone commercialization in order to give them an opportunity to catch up to RCA. The commission sided with RCA's competitors and waited until 1941 to establish transmission standards and authorize full commercial development. Government officials made a decision only after the engineering controversy also became a contentious public dispute.

Chapter 4 analyzes the rise of FM radio during the 1930s and 1940s, focusing specifically on the relationship between the technical decisions of the FCC and controversial social, political, and economic issues. The development of FM radio by Edwin Howard Armstrong seemed to provide a broadcasting system superior to AM; frequency modulation eliminated most of the static that interfered with the reception of AM stations, and FM broadcasting could also provide high-fidelity sound. Despite what its proponents believed was its inherent "technical superiority," FM did not, however, expand to challenge AM broadcasting until the 1960s and 1970s. The decisions made by the communications commission, during the 1930s and 1940s, about such technical matters as the appropriate frequency allocation helped shape FM's growth.

Chapter 5 analyzes the involvement of the FCC in the commercial development of monochrome television during the decade after World War II. The specific focus is an analysis of the controversy over frequency allocation and channel assignments that culminated in the "television freeze" of 1948-52. During this four-year period, the commission refused to consider new applications, pending a review of earlier assignment policies that had led to intolerable co-channel interference. One major policy change after the lifting of the freeze was the authorization of channels in the UHF spectrum (above 470 MHz) for commercial television broadcasting and the reduction of the number of channels in the lower VHF bands (in the vicinity of 44 MHz to 210 MHz) assigned to large cities. The decision had an important impact on the television industry, mainly by hurting the position of the new network, the American Broadcasting Company (ABC). When only two stations were authorized for a particular city, the local stations usually chose affiliation with the two well-established networks, the Columbia Broadcasting System (CBS) and the National Broadcasting Company (NBC).

Chapter 6 examines the involvement of the Federal Communications Commission in the commercial development of color television during the decade after World War II. In 1946, CBS tried to convince the commission to authorize its "field-sequential color system." Although this attempt failed, four years later the commission rejected a system under development by RCA and adopted CBS's color system, despite its incompatibility with monochrome receivers. Then, in 1953, the commission reversed its earlier ruling and chose another system, which the National Television System Committee had developed and which was supported by RCA and other industry leaders. This chapter thus again examines how a major policy dispute involving technical standards was resolved.

Finally, an epilogue relates these earlier decisions about radio and television broadcasting to regulatory developments beginning in the 1960s with the introduction of cable television and continuing through the 1980s and 1990s with innovations in high-definition/digital television.

This book, using detailed analyses of important decisions affecting broadcasting, investigates the involvement of various individuals, groups, and institutions that have sought to shape the industry by influencing policy and decision making. Important participants whose involvement I explore include commissioners and other government officials, members of Congress, engineers in both government and industry, legal experts, station owners, educators, network and advertising executives, members of trade associations, manufacturers, and consumers.[2] In order to keep the work manageable, my analysis focuses on the most influential individuals in the policy arena, but I try to provide a balanced perspective by taking into account the diverse range of interests. Where appropriate, the analysis includes the protests of groups who opposed the decisions of policy makers.

A central theme in the book that provides a framework for analyzing the public-policy debates and negotiations is the tension between technocratic and nontechnocratic views about the role of policy-making institutions. The two different philosophies are especially evident in the contrasting activities and views of Orestes M. Caldwell, who served as a commissioner of the FRC during the late 1920s, and James Lawrence Fly, who served as chairman of the FCC during the late 1930s and early 1940s. Caldwell, who was trained as an engineer, oversaw the development of the first complete allocation, of stations and frequencies, for AM radio broadcasting. This allocation reflected his belief that the government should restrict regulation to the technical aspects of radio. The policy work of the commission, according to Caldwell, was primarily based on "sound engineering principles" and "conditions imposed by . . . stubborn scientific facts."[3] He emphasized that engineers should play the most important role on the commission by constructing a system that efficiently assigned frequencies and station power to the various broadcast stations. The commission, he believed, should deal with problems involving the establishment of standards as narrowly defined—factual issues divorced from social, economic, and political considerations.

In contrast to Caldwell, Fly—a New Deal liberal—believed the commission should not try to restrict its activities to technical issues; he argued for an "integrated and comprehensive regulatory policy" that would also take into account interrelated social, economic, and political concerns. He thought government officials should base their decisions about the authorization of broadcast licenses on such qualitative issues as the character of broadcasters, the use of advertising, the value of large-scale networks, and the educational benefits of programming. Technical experts would still play an important role, but their analysis would not necessarily be limited to narrow technical issues. Unlike later periods in U.S. history, especially after the 1960s, his call for "democratic" evaluation did not necessarily mean encouraging or trying to respond to the grassroots involvement of citizens. Although during these earlier decades there were some exceptions, "elites" in government and industry largely determined national policy.[4] Even when they used consumer surveys, experts played a dominant role in interpreting the results.

Other scholars have identified the tendency toward technocratic thinking as a central feature of twentieth-century liberalism and corporate capitalism.[5] The faith in technical experts has provided an ideological framework for managing the conflicting demands of different groups. It has also served as a foundation for a new managerial and organizational division that has helped bind the federal government to the interests of private industry. The main guideline for the regulation of broadcasting developed by Hoover and established as law in the 1927 and 1934 communications acts was that broadcasters should serve the public interest. But Congress did not explicitly define this standard. For many participants, a technocratic definition seemed appropriate. They reasoned that the use of technical reason would help defuse divisive political debate connected with the many conflicting views of the public good. Neutral and objective scientific experts could transcend contentious disagreements and help identify the public interest. New, complex electronic inventions like radio and television seemed mainly to involve problems of science and technology rather than fundamental issues of private property, democratic rights, or economic control.

This book, through detailed reconstruction of key policy decisions, analyzes the theme of technocracy in the history of broadcasting in the United States. The focus on engineering standards helps illuminate the complex interplay between technical issues and such fundamental concerns as monopoly concentration, patent structure, and control of information. I pay particular attention to the different meanings of *technical* and the way actors established boundaries between different considerations. Although participants seemed to define technical criteria in different ways, the standard view emphasized considerations free from subjective human influence and constrained by physical limitations. Other definitions stressed that the relevant considerations were based on specialized knowledge best understood by experts with specialized training. Policy makers thought they should rely on technical experts not only because the experts were best equipped to evaluate technical facts but also because their professional, scientific training would keep them from making partisan judgments. Although the text of this book sometimes uses the terms *technical* and *nontechnical* when analyzing the involvement of participants, the reader should not interpret this usage as implying the existence of a sharp and definite distinction. I emphasize that these terms are problematic, but the limitations of language cannot always be avoided.

Reflecting the historical complexities I want to explore, my use of *technocratic* also includes a number of interrelated meanings. By technocracy, I refer not only to the belief that technical experts should make traditional political decisions in a society increasingly dependent on complex technologies and technological systems but also to the tendency by decision makers to follow values of instrumental rationality, efficiency, and specialization—values that scholars have identified as a crucial aspect of modern technological society. A commitment to technocratic values has led participants to argue that technological, rather than traditional, political, solutions are most appropriate for solving social problems resulting from technological developments. The use of the concept helps make sense of the tendency of actors involved in complex, hybrid decision making to legitimate their work by arguing that it was simply based on narrow, technical criteria. The examples in this book illustrate how, in many cases, the technocratic ideology did not fit the complex realities of engineering evaluation and policy decision making. Although technical experts employing a strategy of technocratic legitimation sometimes acknowledged privately that, for example, technical and economic factors were interrelated, they generally lacked the critical understanding to recognize how their work might have been shaped by such technocratic values as instrumental rationality and efficiency. Finally, the concept of technocracy also takes into account the policy implications of the commonly held belief that technology is an autonomous force having an internal logic of its own driven by material and quantitative concerns. Historians of technology have used the general terms *autonomous technology* and *technological determinism* to describe these views. In the context of policy making, these beliefs have led people to criticize policy makers for standing in the way of technological progress instead of finding ways to support developments that are almost seen as inevitable. Technology becomes an end in itself instead of a means toward fulfilling intrinsic social, political, and economic goals.[6]

The analysis in this book underscores the importance of pursuing a broad historical analysis, placing technical issues in their proper historical context, and taking into account the different interests and strategies that shape public policy. By analyzing the complex interplay of different factors that went into the formation of public policy for radio and television broadcasting and by taking into account the ideological traditions that framed these controversies, we gain not only a deeper understanding of the institutional and social framework in which policy decisions were made but also a better appreciation of the tensions and conflicts that continue to frame debates involving industrial policy.

I would like to thank the late Hugh Aitken, Bill Aspray, Paul Boyer, Robert Brain, Loren Butler, Susan Douglas, Colleen Dunlavy, Barney Finn, Andrew Goldstein, Jon Harkness, Robert McChesney, Milton Mueller, Rik Nebeker, Eric Schatzberg, Amy Schutt, Emily Thompson, and especially Paul Israel, Ronald Kline, Ronald Numbers, Robert Post, John Servos, and John Staudenmaier for their comments on earlier sections of the book. I am also particularly grateful to the numerous archivists and librarians that facilitated research in very large collections. Versions of this work were presented at the University of Pennsylvania, Iowa State University, the New Jersey Institute of Technology, Rutgers University, George Mason University, and the annual meeting of the History of Science Society. I would like to thank the audiences at these talks as well as Robert J. Brugger, Juliana McCarthy, Dennis Marshall, and other staff at the Johns Hopkins University Press for their valuable expertise.

Work on this book was supported by fellowships from the National Museum of American History of the Smithsonian Institution, the Dibner Visiting Historian Program of the History of Science Society, and the Center for the History of Electrical Engineering, as well as by two grants from the National Science Foundation (SBER-9511607 and SES-9810214). The importance of this support cannot be overemphasized. Finally, I would like to thank Allan Brandt, Peter Galison, Everett Mendelsohn, and all the members of the Department of the History of Science at Harvard University for their generous hospitality in sponsoring me as a visiting scholar. They provided an ideal environment for finishing the final stages of this project.

)))

RADIO
AND TELEVISION
REGULATION

1)))

Engineering Public Policy for Radio

Herbert Hoover, the Department of Commerce, and the Broadcast Boom, 1900–1927

I do not believe any other generation in history has had the privilege of witnessing the progress from birth to adolescence of a discovery so profoundly affecting the social and economic life of the peoples of the world. . . . No other invention in all time invaded the home so rapidly and entrenched itself so securely as radio, and though it is still far from maturity, we see great advances every year.

Secretary of Commerce Herbert Hoover,
September 12, 1925

During the months soon after the Westinghouse Company's establishment of KDKA in Pittsburgh, a boom in radio swept the nation. At the beginning of 1922, 28 licensed stations were broadcasting in the United States to the public; by 1923 the total had risen to more than 550.[1] Earlier, a number of amateur stations had attempted to use radio for public broadcasting, but the first transmissions of KDKA in November 1920, announcing the presidential election returns, proved to be a dramatic new development.

The enthusiasm for radio led many commentators to predict a utopian future for the new public technology. They believed radio broadcasting would raise the cultural standards of the nation and help forge new social and political bonds. As one observer declared: "How fine is the texture of the web that radio is even now spinning! It is achieving the task of making us feel together, think together, live together." Secretary of Commerce Herbert Hoover praised radio's "dawn glowing with the promise of profound influence on public education and public welfare."[2] But to ensure the full beneficial impact of this "profound influence," the country had to address a number of important questions. What would be the relationship between the new use for radio and the established traditions of wireless telegraphy (transmission of coded signals) and telephony (transmission of voice), which private companies and government institutions, especially the U.S. Navy, mainly controlled? How would the country evaluate competing claims of access to the radio spectrum? How would radio broadcasting be supported economically and what role would government—especially the federal government—have in the new industry? The need to attempt to answer these and related questions became more pressing as interference among a growing number of stations broadcasting on a limited range of frequencies threatened to create chaos on the airwaves.

During the early 1920s, the Department of Commerce, under Herbert Hoover's leadership, stepped in to try to manage radio and maximize its potential benefits. Congress had previously placed authority over the regulation of wireless telegraphy and telephony in the hands of the secretary of commerce. Unlike these earlier uses of wireless—or *radio,* the term that became dominant by the early 1920s—transmissions from radio broadcasters were not directed from one point to another (for instance, private messages from a coastal station to a ship at sea) but were specifically broadcast to all appropriate receivers owned by the general public. Despite this difference, Hoover based his efforts to regulate the new technology on earlier legal precedents. He also pointed out that the Department of Commerce was not attempting to impose regulation on the industry, but was responding to demands made by the users themselves. According to Hoover, "this is indeed the only industry I know of which has generally with one acclaim welcomed and prayed for Government control."[3]

Hoover further emphasized that because the regulation of radio broadcasting involved highly technical issues dependent on complex engineering and scientific principles, it presented the federal government with a unique set of difficulties. "The problems involved in Government regulation of radio," he declared, "are the most complex and technical that have yet confronted Congress." Not surprisingly, then, during the 1920s public policy on radio was not simply formulated by politicians and bureaucrats; radio engineers, especially employees of the federal government and members of the Institute of Radio Engineers, played an essential role. This chapter analyzes their involvement and explores that theme in the context of the major tension that emerged between technocratic and nontechnocratic perspectives. Understanding the important role of this tension in the negotiations among the different individuals and institutions working to shape radio broadcasting is fundamentally important. Before pursuing this analysis for the period of the 1920s, however, we first need to recognize that important precedents were established during the decades before KDKA and the rise of radio broadcasting.[4]

The Early History of Wireless

The major technical development that provided a foundation for radio broadcasting was Heinrich Hertz's experimental verification in 1887 of the wave structure of electromagnetic radiation as predicted by James Clerk Maxwell's mathematical equations. The Italian inventor Marchese Guglielmo Marconi most fully explored the commercial possibilities of Hertz's discovery beginning in the 1890s. By developing and improving transmitters, receivers, and antennas, Marconi created a complete system for long-distance wireless communication. After moving to England, Marconi helped set up a private wireless company, which used his system to specialize in point-to-point communication for the shipping industry. He first successfully transmitted wireless telegraph signals across the Atlantic in 1901. Marconi's company soon gained a near monopoly in wireless communications. By 1912, the U.S. subsidiary of the Marconi Wireless Telegraph Company, the Marconi Company of America, controlled nearly all civilian maritime wireless communications from shore stations in the United States and handled most of the nation's other commercial wireless traffic.[5]

During the half dozen years before World War I, two major trends worked against the Marconi Company. Perhaps most important, company officials held to the older "spark technology," while other companies—notably American Telephone and Telegraph (AT&T) and General Electric (GE)—were developing and gaining control of key patents for "continuous-wave technology." Spark transmitters generated radio frequency signals as byproducts of electromagnetic sparking across induction coils. The resulting transmissions, however, produced damped electromagnetic waves of different frequencies. Tuning to onc frequency was difficult, interference among transmitters was a major problem, and the technology was not entirely satisfactory for voice transmission, or telephony. AT&T and GE acquired control of two new inventions that became the basis for continuous-wave transmissions: the alternator and the audion (or triode vacuum tube). AT&T secured patent rights to the audion from the inventor Lee de Forest; GE gained control of patents on the alternator developed by its employee Ernst Alexanderson. Continuous-wave technology produced high-power signals of constant frequency, which stations could more easily use to transmit the human voice.

The second trend working against the Marconi Company was the growing influence of the military, especially the navy, on the wireless business in the United States. Key navy officials—especially Josephus Daniels, secretary of the navy from 1913 to 1921—advocated complete naval control of wireless. Daniels was especially interested in keeping U.S. technology out of the hands of foreign companies. Key officers worked closely with American companies to integrate the new continuous-wave technology into all aspects of naval operations.[6]

During World War I, the military services did succeed in assuming control over wireless, in the name of national defense. In this case, the record of government control was generally good: military demands for improved apparatus and a government-supported patent moratorium that promoted innovation supported new research and development. The great potential of the vacuum tube as both a detector and a generator of radio waves was realized during wartime. However, other cases of government control during the war, especially of public utilities and the railroads, were far less successful. Because of these experiences, the public was not prepared to support Secretary Daniels's request that Congress authorize a continuation of naval control of wireless after the war. The newly elected Republican Congress exploited this public sentiment against a continuation of a wartime policy sponsored by a Democratic Congress. Although Daniels was forced to give up on his primary goal, he continued to pursue a secondary goal of preventing foreign control of American wireless technology during the postwar period. The navy had confiscated American Marconi's long-distance shore stations during the war and was anxious to find a way to avoid returning them. The navy was also concerned about GE's arrangement after the war that would have given American Marconi de facto exclusive rights to its alternators. In response, during the summer of 1919, the navy convinced GE to help establish a new American company, the Radio Corporation of America (RCA), formed from the acquisition of American Marconi. Using apparatus produced by GE and other U.S. manufacturers, the new all-American company retained American Marconi's monopoly of long-distance point-to-point service.[7]

RCA also supported the expansion of radio broadcasting during the 1920s. But the military influence on the institution was fundamental. The wartime experience had demonstrated the benefits of both monopolistic control and the suspension of competing and contentious patent claims. Within two years after the establishment of RCA, a series of agreements was worked out among RCA and the other major companies involved in radio. No one business controlled patents to a complete technological system of continuous-wave transmission and reception. As a result, the companies holding major patents—RCA, AT&T, GE, Westinghouse, and the United Fruit Company—agreed to extensive cross-licensing arrangements. The companies also consented to divide all aspects of the radio business. RCA retained exclusive rights to international wireless telegraphy and nonexclusive rights to international telephony. AT&T retained its control of most wireless telephony. GE and Westinghouse would manufacture radio receivers and radiotelegraphic equipment; AT&T (through its subsidiary Western Electric) would control the manufacture of wireless-telephone transmitters. RCA also agreed to buy from GE 60 percent of the radio apparatus it sold; the other 40 percent would come from Westinghouse. GE, Westinghouse, and AT&T had representatives on the board of directors of RCA and owned stock in the company. As Hugh Aitken pointed out, after the final agreement with Westinghouse in 1921, RCA controlled, "directly or through its affiliated companies, every American patent of importance in the field of continuous wave radio."[8]

Some of the same forces that had come together to help create RCA also shaped the early efforts by the federal government to regulate wireless. Why did the regulation of wireless seem necessary in the United States and in other countries? Originally, the major reason for government intervention was to ensure safety at sea. Distress calls would be ineffective if ships did not carry wireless equipment or maintained incompatible systems using different frequencies, especially in emergency situations. Nations using radio held international conferences at the beginning of the century to deal with these issues. In most countries, the central government assumed complete control over the radio spectrum. Government ownership seemed necessary because of the crucial military and civil uses for radio. The United States lacked the same traditions of government control, but public opinion also did not favor private ownership of the radio spectrum. This area might be a new continent for exploration, but officials questioned whether the government could divide up something as intangible as the airwaves—they came to be known as "the ether"—into sections of private property the same way it had parceled out land. During the nineteenth century, federal land policy had encouraged citizens to claim public land at minimal or no cost and transform it into private property. But Progressive-era politicians concerned with the public interest argued that the spectrum was different; they feared that if the government allowed property rights, one group might end up with a monopoly of ideas and information and the ability profoundly to shape public opinion. Thus, radio policy in the United States was grounded in the conviction that the spectrum belonged to the public. Everyone should have a right to obtain a license and use the spectrum. However, especially after the rise of radio broadcasting during the 1920s, policy makers increasingly viewed the radio spectrum as a finite resource. At any one time, only a limited band of frequencies was available for wireless, and interference among stations (often using poorly tuned equipment) limited the number that could transmit at any one time. All citizens might own the ether, but if everyone tried to use it its value would be destroyed. Throughout the early history of radio (at least until 1927), radio policy in the United States had to deal with a potential contradiction. Decision makers wanted everyone to have a right to use the spectrum, but they increasingly came to the conclusion that the government would have to place limits on access to the radio spectrum to avoid overexploitation or, in other words, destructive interference.[9]

Congress was not convinced of the need for legislation until a shipping accident in 1909 demonstrated the value of wireless for safety at sea in a spectacular way. Maritime officials praised a single wireless operator for saving the lives of twelve hundred people. The 1910 Wireless Ship Act mandated that the government give priority of access to the spectrum to operations aimed at ensuring public safety. The law required that most oceangoing steamers have a skilled wireless officer and a wireless apparatus capable of communicating with any other system located within a radius of one hundred miles.[10]

But the 1910 law did not help alleviate the problem of interference; in fact, by expanding the number of users of the spectrum, Congress probably inadvertently made things worse. Most interference was unintentional, caused by a large number of closely spaced stations, many using "dirty" transmitters producing spurious signals. Some interference, however, was intentional—and when it occurred, amateur operators were usually blamed. In addition to the navy and private companies, the amateurs were the third major group using wireless before 1920. Amateur operators included a large number of boys and young men who shared a hobby of communicating using homemade equipment. The introduction of the crystal detector in 1906 helped support this democratization of wireless. Amateur operators provided an important early audience for radio broadcasting; they also made important experimental broadcasts of music and entertainment, many years before the establishment of KDKA by Westinghouse. The number of amateur stations operating in the United States before World War I is unclear; in 1912 the *New York Times* estimated that several hundred thousand existed. The amateurs tended to view the spectrum as a new, wide-open frontier, akin to the American West, where men could pursue individual interests free from repressive authoritarian and hierarchical institutions. They resented attempts by the navy and private companies to monopolize the spectrum for commercial or military gain. This antiauthoritarian sentiment led a few amateurs to intentionally transmit false or obscene messages, especially to naval stations. The U.S. Navy complained bitterly about amateurs sending out fake distress calls or posing as naval commanders and sending ships on fraudulent missions. Josephus Daniels and other naval officers used this threat to national security and safety as a justification for seeking total naval control of wireless.[11]

The perceived need to discipline amateurs in order to reduce interference led Congress to begin to consider legislation more sweeping than the 1910 Wireless Act. During that same year, Congress considered six different proposals for new legislation. But it took a new tragedy, in April 1912, involving both issues of public safety and interference caused by amateurs to convince Congress to pass comprehensive legislation. The event was the sinking of the *Titanic*, with the loss of more than fifteen hundred lives. Citizens were horrified to learn that two of the ships closest to the *Titanic* had not been able to respond to the radio distress call; in one ship, the wireless operator was asleep; in the other, no wireless equipment had ever been issued. Politicians responded to the public outcry by condemning the 1910 Wireless Ship Act as inadequate. Even more shocking was the revelation that constant interference and false messages from malicious operators had hampered the rescue effort dispatched to help the *Titanic*. The press blamed the amateurs, who lost even more credibility.[12]

Four months after the *Titanic* disaster and in order to comply with an international convention enacted that same year in London, Congress passed comprehensive legislation regulating the use of radio, the Radio Act of 1912. It remained the only law of its kind in the United States, despite more than thirty attempts to introduce new legislation, until it was revised in 1927. The 1912 act required that the Department of Commerce license all radio operators. The department, which already had limited authority under the 1910 act, was authorized to make necessary frequency changes when private stations interfered with military transmissions. The law also established stringent requirements that ships have at least two radio operators and maintain superior "clean" wireless equipment that would not cause spurious interference. Radio operators had to give any station making a distress call priority of use of the spectrum; interference had to be avoided. Following international agreement, U.S. citizens were required to set aside the 300-meter (999.4 kHz) band for emergency transmissions. In the event of war, the statute authorized the military services to take control of all private stations. Finally, the legislation divided up the use of the spectrum by assigning specific frequencies to different groups.[13]

The new allocation scheme was consistent with international agreements already being followed in Europe. It reserved frequencies between 187.4 and 499.7 kHz for the federal government, mainly the U.S. Navy. Private stations were given the use of frequencies above 499.7 kHz and below 187.4 kHz. The allocation relocated the amateurs to the shortwave region above 1,500 kHz, a band not considered usable at that time. Thus, the 1912 Radio Act implicitly clarified the criteria that the federal government would use in judging which users of radio should have priority of access to the spectrum. As Susan Douglas argued, "what established merit in 1912 was capital investment or military defense, coupled with language that justified custodial claims based on invaluable service to humanity." The act did not give authority to the secretary of commerce to deny a license to any individual; it therefore upheld the conviction that since the spectrum belonged to the people, everyone should have a right to obtain a license. But some parts of the spectrum were more desirable than others; by placing amateurs in an undesirable section, Congress was effectively making a decision about limiting access to the use of radio. A decision that seemed to be purely technical in nature had significant economic and social dimensions.[14]

In the public debates over national radio policy that occurred before the first broadcasts of KDKA, an important theme emerged that would play a crucial role in the efforts by Hoover and others to interpret and administer the 1912 Radio Act for public broadcasting. A number of individuals, especially engineers and business leaders, argued that national radio policy be guided by technical considerations evaluated by technical experts. This technocratic position seemed appropriate since the regulation of radio was driven by the technical problem of interference, which in turn partly resulted from the technical limitation of a finite spectrum.

Some congressmen were convinced by the testimony of engineers and scientists that new radio legislation was unnecessary because engineering solutions to interference were just around the corner. Specifically, they promised that the radio industry was on the verge of developing "clean" transmitters and other new apparatus that would produce sharply defined signals and allow a growing number of stations to fit into the band of available radio frequencies. During congressional hearings in 1917, Alfred Goldsmith—professor of physics at the College of the City of New York—testified against a bill that proposed naval control of wireless as a naval solution to interference, by assuring members of Congress that "the problem of interference is sure to be solved in the near future by technical means now under development by the companies." Michael Pupin, professor of physics at Columbia University and an important inventor of components for electrical communications, also reassured Congress that "things are being done today by well organized industrial research laboratories which will undoubtedly lead to wonderful results so far as preventing interference produced by the acts of man are concerned." The engineers and scientists who testified against a naval monopoly believed that legislation or government control would only stifle research. Pupin even went so far as to argue that the technical problem of interference be seen as a positive challenge that would stimulate technological development. "If I had my own way," he declared, "I should produce as many interferences as I possibly could, for the purposes of development of the art."[15] The scientists and engineers testifying before Congress believed they deserved a special role in advising the country on national radio policy. Radio was their invention and they felt confident future research would assure its great promise. Their testimony also implicitly demonstrated a commitment to particular economic and social views: technical progress should not be stifled by government control but should be driven by the industrial research laboratories of GE, AT&T, Westinghouse, and other large manufacturers.

In retrospect, it seems clear that the predictions of these technical experts were not entirely realistic. The introduction of vacuum-tube technology after World War I did lead to the widespread use of high-quality continuous-wave transmitters and tunable receivers. But sometimes new advances created their own problems. One popular receiver using vacuum tubes could actually become a transmitter if improperly adjusted; this resulted in thousands of new sources of potential interference. Irrespective of this new problem, improvements in the sensitivity and selectivity of transmitters and receivers that were possible at that time would not have been enough to overcome the severe problem of interference and spectrum scarcity that developed after the rise of radio broadcasting during the 1920s. As Aitken argued, given the decision not to limit access to the broadcast spectrum by authorizing private property rights, "technological advance alone would not have solved" the problem: "There were too many beasts foraging in the pasture."[16] But the technocratic arguments helped defeat naval attempts to gain control of radio and, as we will see, continued to play an important role during the policy debates of the 1920s.

The Department of Commerce Takes Control

When radio broadcasting emerged during the early 1920s, it upset the balance of power among different groups of radio users in the United States. Broadcast stations competed with government institutions and private companies specializing in point-to-point transmissions. An intragovernmental contest also complicated matters tremendously. The navy had been unsuccessful in its bid to gain complete control of radio after World War I, but it continued to seek to influence policy, especially by trying to maintain close contacts with RCA. At the same time, the Post Office Department resisted naval influence and, following the pattern in European countries, attempted to assume control of all communications. In 1919, by authorizing construction of a series of land radio stations to support the new airmail service, Congress affirmed that the Post Office Department would have an important role to play. Other stations, managed jointly with the Department of Agriculture, transmitted market and weather reports to the public. While government institutions competed for influence, private companies resisted all attempts at government control. During the first year after its establishment in 1919, RCA and the navy cooperated on policy matters; however, when RCA, by constructing its own coastal stations, began during 1920 and 1921 to threaten the navy's dominance of maritime traffic, the cordial relationship ended. RCA and the other commercial companies turned to Secretary of Commerce Hoover for support; he championed their cause and used it to gain the upper hand in the intragovernmental contest.[17]

By working closely with industry representatives, Hoover forged an essential link between government and private enterprise. But engineers played the key mediating role, serving to cement a relationship that otherwise would have been fragile. Recent scholarship has identified technical expertise and technocratic values as essential elements of twentieth-century "corporate liberalism." A reliance on technical expertise helped overcome potential contradictions by making manifest an objectified, neutral public interest. Even before Hoover became secretary in 1921, the Department of Commerce had sought to take advantage of engineers from outside the government to advise the department on policy decisions that would affect the industry.[18]

A major policy issue during the period following World War I but preceding the rise of radio broadcasting was a reevaluation of international agreements involving the use of the radio spectrum. An Interallied Wireless Commission met in Paris after the war and issued recommendations updating the preceding international radio convention, held in London in 1912. The military representatives of the commission asked the U.S. government to approve proposed changes so they could be incorporated into the "Treaty of Peace and impressed on Germany," at least until the next international radio convention. The Departments of the Army, Navy, and Post Office approved the proposals, but the Department of Commerce supported industry representatives who criticized the new rules for favoring the military. The existence of major divisions within the government, not only between officials from private industry and the military but also among the different private interests, became obvious during a meeting sponsored by the Department of Commerce on March 30, 1920, in which military representatives explained the draft proposals to industry representatives. The commercial concerns generally agreed that the proposals were too rigid; they were especially upset that government officials wanted to set aside a large block of frequencies exclusively for military use. However, the companies disagreed on specifics. For example, AT&T wanted more channels assigned for the use of radio telephony. Small, independent manufacturers complained that a proposal to outlaw spark transmitters from wavelengths of more than 1,500 meters would favor RCA and the other members of the "radio trust" who controlled the patents to continuous-wave apparatus. Recognizing that they would not be able to come to any definite agreement on evaluating the proposals, the participants at the Department of Commerce meeting, who were in many cases employed as engineers, agreed to assign a committee of engineers the task of evaluating the technical issues and advising the department on how to proceed. The representative for AT&T, J. J. Carty, argued that a committee of technical experts would give "the best answer for all parties concerned, not for one particular group but the best balanced judgment of all of the interests." The Army representative agreed that a technical committee would be able to "harmonize the various" interests.[19]

The discussion leading to the organization of an advisory committee underscored important tensions and conflicting views about the use of technical experts for policy making. The obvious candidate for the advisory committee was the Wave Length Allocation Committee of the Institute of Radio Engineers. The institute, which had been founded in 1912 to advance the field of radio engineering and enhance the status of professional radio engineers, sought to encourage technical developments and construct engineering standards, as well as, especially during the 1920s, influence public policy and direct industrial growth. By World War I, large corporations had become the major employers of radio engineers. The tension between a commitment to the interests of the companies who paid their salaries and an obligation to independent technical evaluation became increasingly important as leaders of the institute began to play a major role advising the government about regulating radio. The institute had established the Wave Length Allocation Committee before the war to help formulate radio policy; some of its members participated in the March 30 meeting held by the Department of Commerce. Government officials placed the institute in charge of organizing the advisory committee that would evaluate the proposed changes to the 1912 convention. Because of its outward commitment to the "general welfare of radio communication" (that is, not to any one particular commercial interest), the professional society seemed the obvious candidate to successfully "harmonize" the different interests. The president of the institute, John V. L. Hogan, assured Secretary of Commerce J. W. Alexander that it would "find it impossible to express in a Committee Report the views of any corporate or other radio interests as distinguished from recommendations conducive to the best progress of the radio arts and practice." The AT&T official at the March 30 meeting confirmed the special role of the Institute of Radio Engineers. Carty related how he had told one of his engineers who planned to serve on the Wave Length Allocation Committee that "you are to act as a member of that body on your own engineering conscience and what that imposes is the thing that you must do."[20]

But Secretary of Commerce Alexander and some of the other participants at the March 30 meeting emphasized that they did not think the Wave Length Allocation Committee could satisfactorily represent all the different radio interests. He thought it would be "unwise to refer a subject as that we are considering exclusively to one private organization, even though of such high standing as the Institute of Radio Engineers." The secretary questioned whether the institute's commitment to unbiased pure science was really what the department wanted: "To the Department of which I am head Science is the handmaid, not the mistress of Commerce, and I trust the committee will bear this in mind. I am a little afraid you are raising academic questions, for the men of the conference impressed me as practical men and such men usually get results. If any irreconcilable conflict should arise in the committee between Commerce and Science probably you can ease your consciences by making one report as scientists and another report as men of business."[21]

Industry representatives at the March 30 conference noted not only that the committee of the Institute of Radio Engineers did not include engineers representing key commercial interests, but, as they pointed out to Alexander, the Department of Commerce, specifically its major division, the Bureau of Standards, was not "represented as actively as it should like to be." As we will see, important engineers in this bureau would help link government and private industry. The Bureau of Standards pursued radio research needed to administer the 1912 Radio Act. The critical sentiment against using the Wave Length Committee led the participants at the March 30 conference to expand its membership to include more representatives from industry and the government. Presumably, members of the new committee could make decisions that were based not only on their "engineering consciences" but also on their employers' interests. Although the new committee was not one "of the institute," it still maintained informal contacts; of the new members added to the original institute committee, all but one (the post office representative) were engineers who were also members of the institute. The secretary of the institute, Alfred Goldsmith, served as chairman of the new committee—the Department of Commerce Radio Conference

Committee. Despite the president's earlier insistence that the committee did not need new representatives because all institute members were expected to set aside their private interests, Hogan did acknowledge to the secretary of commerce—before he understood that the new committee would be formally separate from the institute—that "institute committees invariably attempt to secure the widest diversity of views on the questions in hand, and therefore your suggestions as the appointment of additional members have been cordially welcomed."[22]

Concerns about the implications of international agreements for the radio industry continued to draw private industry and the Department of Commerce together during 1921. Industry officials were especially upset that military representatives persisted in dominating international conferences, including a meeting in Washington in December 1920 and another in Paris in June 1921. Different national officials held meetings to prepare for a planned international communications conference; such a conference did not meet until 1927. The military representatives at the Washington meeting drafted a proposal for a Universal Electrical Communications Union that favored government control by unifying regulation of wireless, wire telegraph, and cables. Industry officials protested that the Washington meeting ignored the modified proposals developed after the Department of Commerce meeting on March 30, 1920; they also believed the unified regulations would hurt wireless and stifle private enterprise. Industry officials met with Department of Commerce representatives in May to plan strategy before the June meeting in Paris of the Interallied Provisional Technical Committee. But military representatives who ignored the interests of private industry also dominated the Paris meeting. Of the participants at the Paris meeting, 75 percent represented military departments of governments.[23]

The disappointing results of the Paris meeting helped solidify the developing connection between private companies and the Department of Commerce. Hoover, as the newly appointed secretary of commerce, initiated a series of meetings, beginning in October 1921, with industry representatives from RCA, AT&T, Western Electric, GE, and other major companies. He wrote Goldsmith on October 6 that "it now becomes necessary to develop plans to give suitable representation to United States commercial interests in the next general world conference on communication." Engineers again played a crucial role in linking Hoover's department to private industry. Goldsmith, the secretary of the Institute of Radio Engineers, and Wesley Stratton, the head of the Bureau of

Standards, organized the meetings initiated by Hoover. Although the institute tried to prevent the interests of individual companies from influencing its activities, the organization supported the idea of private control of radio as opposed to military or Post Office Department domination. Responding to the recommendations of the Washington meeting of December 1920, Goldsmith expressed the sentiment of the institute, as well as his employer RCA, when he declared that "we object to a tendency toward government control which is felt throughout the draft."[24]

Hoover and the First Radio Conference

By the fall of 1921, problems with domestic radio began to overshadow international concerns. The radio broadcasting boom had taken off the previous year; Hoover called it "one of the most astounding things that has come under my observation of American life." In February 1922, he estimated that the number of receivers being used by the public had increased in less than one year from fifty thousand to six hundred thousand. The 1912 legislation, which gave Hoover authority to issue licenses and try to prevent interference, had not taken into account this new use for radio. Government officials had reserved most of the radio spectrum for military and commercial point-to-point wireless transmitters. Hoover managed to set aside one frequency, 832.8 kHz, for the many broadcasters operating in the new category he established on September 10, 1921: "limited commercial stations." He allowed use of the 686.6 kHz channel for stations broadcasting government agricultural and weather reports. The separate class of "broadcast" stations using the 832.8 kHz frequency was formally established in January 1922. Hoover's department sanctioned these stations to broadcast specific kinds of programming—among them news, music, entertainment, speeches, and market reports—that other stations, notably those run by amateurs, could no longer transmit. But in many areas interference became intolerable, not only among broadcasters but also between broadcasting stations and other services, especially maritime radio. A major source, though by no means the only source, of interference was poorly tuned transmitters. Listeners complained that KDKA sometimes covered all the frequencies from 832.8 kHz (where it was supposed to broadcast) down to 428.3 kHz. In order that the "maximum public good shall be secured from the development of this great invention," Hoover sought to gain support for new legislation giving the secretary of commerce clear legal authority to regulate radio broadcasting. In February 1922, he brought together experts in the field to help draft legislation and convince Congress of its necessity. This would be the first of a series of radio conferences Hoover organized during the next four years to assist his efforts to support the growth of the new public technology.[25]

Hoover used the Department of Commerce to build up a governmental apparatus that would assist and guide private industry rather than intrusively regulate. He employed his managerial skills to bring together industry experts willing to work cooperatively with government representatives to arrive at rational solutions to problems. Hoover's *associationalism* (to use Ellis Hawley's term) not only idealized technical expertise but also the closely related institutional structure of professional engineering and trade associations. In the many conferences he helped organize, he relied on individuals from these groups to reach consensus. His engineering background obviously influenced his methods of administration; however, it is important to keep in mind that many administrative theorists during this period who were not engineers shared his desire to place power with nonpartisan experts. Hoover's "use and understanding of technical language," according to John Jordan, "illustrates the persistence of engineering modes of thought in the political discourse of the 1920s."[26]

Because radio depended on "technical" or "scientific" elements, such as the propagation properties of different layers of the ionosphere, to a much higher degree than other industries, Hoover's radio conferences placed an especially strong emphasis on technical expertise. In his opening address at the first conference, he emphasized that "the delegates to this conference were representative scientific and technical men." The major problem that the conference addressed was "what extension in the powers of the department should be requested of Congress in order that the maximum public good shall be secured from the development of this great invention." Conference discussions underscored two general views about the proper role of the secretary of commerce in regulating radio to maximize the public interest. The RCA representative expressed a strong technocratic philosophy that government regulators limit their activities to narrowly defined technical considerations. On one of the most important issues of radio regulation—deciding what criteria to use in the allocation of frequency bands to different services—RCA officials said this was "wholly a technical question" that should be solved through the use of quantitative criteria: "The rule under which we have been endeavoring to operate is that of doing the greatest good to the greatest number."[27]

RCA was responding to other conference participants who recommended that the secretary of commerce use social and economic factors in determining who should have access to different parts of the spectrum. A number of participants argued that the department establish a hierarchy based on standards of "character, quality, and value to the public": the first priority should go to government broadcasting, the second to educational and public broadcasting, the third to private broadcasting, including entertainment and news, and the fourth to toll broadcasting (the sale of airtime to private firms or groups mainly interested in selling products or services). The most important issue for a number of participants was that the government should not promote commercial radio at the expense of educational broadcasting. Cyril M. Jansky, an engineer at the University of Minnesota who helped set up an early educational station, contended that "stations established for the sake of creating a market for apparatus or for advertising purposes should not be granted the privileges at the expense of stations which broadcast educational information where there is no monetary interest involved." The use of advertising to support private broadcasting was still being used only experimentally during this early period. Most conference participants agreed that they should not encourage the practice. Hoover announced in his opening address that "it is inconceivable that we should allow so great a possibility for service, for news, for entertainment, for education, and for vital commercial purposes, to be drowned in advertising chatter." RCA officials did not favor advertising, or at least did not make a statement about it, but they did believe that government officials should not take into account such considerations as commercial versus educational broadcasting when deciding about licensing stations or allocating services to different parts of the spectrum. "All transmitting stations," they believed, "should be subject to the same general laws." The company maintained that the effort to set up a hierarchy based on qualitative standards would only stifle technical progress. Technical progress itself would best dictate its own rules. According to RCA, no effort should be made to "impose upon the radio art detailed statutory rules and regulations. . . . It seems to us that greater advances can be made if for the present at least it is governed in all of its aspects by its own regulation."[28]

As part of his effort to get the conference participants to go beyond technocratic reasoning, Jansky pressured Hoover and the other leaders to include an investigation of RCA's licensing policy in the conference proceedings. Consideration of this issue, Jansky later reported, "was the point of more discussion at the meetings of the conference than any other one thing." He was especially concerned that the conference participants take into account the patent and monopoly implications of recommendations. Although "from an engineering standpoint," regulations eliminating spark transmitters would "be highly desirable" because of the resulting reduction in interference, Jansky believed the decision "would automatically place the sale and control of the use of the necessary equipment for practically all types of radio communication in the hands of" RCA. He also warned that a proposal to place public and private broadcasting in two different frequency bands (968 kHz for private and 200–285.7 kHz for public broadcasters) would benefit private stations because 90 percent of receiving sets would not be able to tune to the shorter frequencies. Jansky called on the conference to include recommendations for new legislation giving the commerce secretary authority to compel owners of patents to adopt open licensing policies, allowing any manufacturer to use patents after paying the necessary fee. To convince the conference to consider his accusations of potential monopoly control seriously, he argued that "in connection with the relative importance and treatment of public and private broadcast stations that at least 80 percent of the broadcast service supplied to the public . . . is supplied from stations owned or controlled by the Radio Corporation of America and affiliated companies." This figure seems an exaggeration, but Jansky's statement does indicate the importance of the monopoly issue and how some independent broadcasters felt threatened by RCA.[29]

Both Hoover and the key congressional adviser at the conference, Congressman Wallace White, a Republican from Maine, decided that it would be "unwise" to undertake an in-depth examination of licensing and monopoly issues. During that same year, the Federal Trade Commission (FTC), partly under pressure from Congress, decided to investigate the charge that RCA was restraining trade through its selling practices. RCA did begin to adopt a more open licensing policy as a result. But during 1922, the conference followed Congressman White's advice and decided to avoid unnecessary conflict by concentrating on "securing the minimum legislation necessary to provide the Secretary of Commerce with sufficient authority to license, refuse to license, or revoke, for cause, a license of a radio transmitting station."[30] Conference participants justified these actions based on the "technical" need to reduce interference; they decided that controversial socioeconomic questions of patent policy and monopoly control were matters that the FTC should investigate.

But the conference's recommendations for new legislation included the provision that the "order of priority of the services be Government, Public, Private, Toll." It also recommended that Hoover take into account the quality and public interest of commercial stations when evaluating licenses and allocating frequencies. Another factor that would "affect privileges extended to" stations was "the sharpness of the emitted wave." The conference wanted the government to discourage poor-quality equipment that caused excessive interference. And it recommended that the government "absolutely" prohibit direct advertising. The only proper advertising should be "limited to a statement of the call letters of the station and of the name of the concern responsible for the matter broadcasted." By making a distinction between direct selling or pricing of products and the more indirect methods of sponsored programming, the conference made limited advertising palatable. Finally, the conference affirmed the idea that the radio spectrum was a public resource; stations could not claim to own the frequencies they were allowed to use.[31]

The radio conference helped solidify the Department of Commerce's role as the major government institution that would oversee radio regulation. Participation by the Post Office Department and the U.S. Navy was restricted. After the conference, in an effort to end the interdepartmental contest, Hoover helped organize a new organization to coordinate all government interests in radio, the Interdepartmental Advisory Committee on Government Broadcasting. To avoid Post Office Department control, the navy supported expanded responsibilities for the advisory committee, which in January 1923 was renamed the Interdepartment Radio Advisory Committee. The new government committee acknowledged the importance of private radio, independent of military control (except during war); in February, the committee transferred all broadcasts of the Department of Agriculture from the post office to the navy. The naval compromise helped defuse the interdepartmental dispute and ended the ambitions of the Post Office Department. Late in 1923, James G. Harbord, the president of RCA, spoke assuredly of the end of the "heresy of government ownership, especially in radio matters." The navy would control its own stations; the Department of Commerce would regulate radio broadcasting, both private and public.[32]

Legislative Defeats and Conference Proceedings

Following the first radio conference, new legislation based on the conference recommendations was introduced in Congress. Congressman White and Senator Frank Kellogg, a Minnesota Republican, submitted identical bills during 1921. The new legislation sought to replace the 1912 Radio Act, mainly by giving the secretary of commerce extensive authority to regulate radio. The two bills also recommended the establishment of a committee of technical experts to advise the secretary. This legislation, as well as similar measures introduced by White during 1923, did not make it through Congress. People on all sides of the issue had objections. Officials at RCA argued that before the government stepped in to regulate, the industry needed to gain a solid economic base. Both AT&T and RCA objected to a "monopoly clause" in White's legislation that prohibited the secretary of commerce from assigning a license to a station violating antimonopoly principles. Goldsmith, in his capacity as a representative of the Institute of Radio Engineers, complained that the monopoly clause was "entirely out of order." White's legislation included a provision establishing a commission of twelve experts, six government and six nongovernment representatives, to advise the secretary of commerce about radio regulation. Members of the Department of Commerce recommended that the nongovernment members have the highest level of competence "not only from an engineering point, but from a business and general public service viewpoint as well." The close alliance between the department and the major companies involved in broadcasting was evident in the department's judgment of the best candidates. One of the department's radio inspectors doubted "very much if this type of a committeeman can be procured without going into the field of some of our big commercial enterprises." The department considered Goldsmith an ideal candidate for the commission because he would not only represent the institute but also RCA and GE, the two companies paying his salary as a research engineer. Goldsmith had "hinted" to Department of Commerce officials that "he would like to serve as a member even without compensation," giving as a reason his "duty" to the "American public." Key professional engineers like Goldsmith thus played a crucial role forging a link between government and private enterprise and defining the public interest in technical terms supporting industry development.[33]

Hoover's legal authority to regulate radio broadcasting without new legislation was highly questionable, especially after judicial decisions during the early 1920s. On May 23, 1921, the department revoked the license of the Intercity Radio Company, charging it with flagrant interference of government broadcasts. The Court of Appeals for the District of Columbia Circuit ruled against Hoover in November 1921, arguing that he could not deny a license to an applicant. The company went out of business before the U.S. Supreme Court had a chance to rule, but this and other lower courts' rulings demonstrated that Hoover's authority under the 1912 act was clearly problematic.[34]

Despite judicial and legislative defeats, Hoover proceeded as if he did have clear legal authority to regulate radio broadcasting, partly by pointing to the clause in the 1912 act that gave the Department of Commerce responsibility for "preventing or minimizing interference." To avoid the controversial action of denying licenses to stations, the Department of Commerce, following the first radio conference, had attempted to develop a rational allocation system, which mainly involved having nearby stations broadcast at different times. When the general broadcast frequency of 832.8 kHz became congested during 1922, even with complicated time-sharing arrangements, the department set aside another frequency for a different category of station. Officials allowed these "class B" stations, first authorized in August 1922, to broadcast at higher power levels (between 500 and 1,000 watts) on a frequency of 749.6 kHz, but Department of Commerce experts expected them to maintain "high-quality" programming as well as equipment able to stay well-tuned to the required frequency. Although Hoover publicly opposed any action that might be construed as censorship, radio inspectors did not, in practice, permit class B stations to use "mechanical music, phonographs, and things of that kind." Within a few months, the new frequency for class B stations also became congested, even after the department instituted time-sharing between stations.[35]

By early in 1923, Department of Commerce officials realized that broadcasters needed new channels in order to alleviate the problem of interference. Since the first conference, the number of broadcast stations had risen from 60 to more than 550. The department was especially interested in allocating to radio broadcasting the band of frequencies (187.4 kHz to 499.7 kHz) that the 1912 Radio Act had set aside for use by the government, mainly the navy. To gain public support for this and other administrative measures to reduce interference, Hoover organized a second radio conference in March 1923. Conference participants, including eight government and eleven nongovernment representatives, agreed on the importance of the proposal to clear a band of frequencies for broadcasting. They also approved Hoover's policy of establishing two classes of stations (high-power and low-power) and decided that they should be located in different frequency bands. Despite seeming to violate the 1912 legislation, the recommendation to use the government band became policy because government representatives agreed not to oppose the move. With the opening up of new frequencies higher in the spectrum, the lower government band was no longer of crucial importance, especially to the navy. The navy also realized that it would benefit from the decision because it could justify appropriations for new, better-quality apparatus, necessary for transmitting in the higher frequencies.[36]

At the second conference, as at the first, engineers again played an important role. Their cultural authority helped buttress the conference's opinion that the secretary of commerce did indeed have legal authority to implement broad policy recommendations, including rearranging frequencies assigned to different users of radio and denying licenses to broadcast stations that caused excessive interference. Goldsmith, representing the Institute of Radio Engineers, first introduced the latter opinion at the conference. Although the conference was mainly concerned with solving the "technical" problem of interference, the engineers providing advice thus also made important "policy" recommendations. Goldsmith urged the conference to convince amateurs to avoid using radio during periods when stations broadcast religious programs; this, he said, would "ensure the reception of such religious services in any given locality." Further, testimony from John Hogan, the institute president, emphasized that some of the conference's technical recommendations, for example the establishment of different classes of stations based on transmitter power, implicitly involved nontechnical considerations. He assumed that the programming of "local interest" broadcast by low-power stations would be qualitatively inferior to the "high grade" programming of high-power stations.[37]

After agreeing to continue to allow broadcasts on one of the three channels previously set aside for broadcasting, Hoover authorized three classes of stations on May 15, 1923: class A stations, assigned to operate at no more than 500 watts in the band from 999.4 to 1365 kHz; class B stations, authorized to use between 500 and 1000 watts in the bands from 550 to 800 and 870 to 999.4 kHz; and low-power, class C stations, using the old 832.8 kHz channel. Department of Commerce radio inspectors reassigned stations to the new frequencies after the May decision. Particular regions, especially major metropolitan areas, were given specific frequencies. For example, New York City received three frequencies for its class B stations. Before instituting the new allocation scheme, the department warned stations causing excessive interference that they might loose their licenses if their offenses did not stop; however, officials apparently never followed through with these threats. Despite efforts to develop a smoothly functioning allocation system, interference and time-sharing problems continued into 1924.[38]

Third Radio Conference: Superpower and Interconnection

Following further legislative defeats and new sources of interference, Hoover held a third radio conference in October 1924. To gain the support of all users of radio, he increased the number of delegates to ninety—including not only engineers but also business leaders, government officials, and individuals representing the broadcast listeners. Eight subcommittees met to discuss the major problems of radio. One of the conference's principal tasks was again to address the problem of reducing interference between stations. But two new developments, interconnection and superpower broadcasting, introduced new dimensions to this old problem.[39]

Since the second (March 1923) radio conference, the large firms involved in communications had begun to perfect these new techniques in order to provide nationwide radio service and bring programming of "larger centers of art, music, and events of public interest to the more remote" regions. A number of earlier technological innovations provided important precedents for the development of national radio service. The large companies in the radio industry believed decisions to build centralized, national systems in the telephone, telegraph, and electric-lighting industries had resulted in reduced costs and increased efficiency for businesses as well as better service to consumers. The economic structures and organizational forms of these older technological systems provided crucial models for thinking about and organizing radio broadcasting. Corporate executives viewed radio broadcasting as the latest in a series of progressive developments in the history of technology. The sense of inevitability assumed by this progressive logic seemed to indicate that national radio systems would naturally evolve along the lines of these earlier precedents. A tradition that stressed the importance of new communications technologies in helping the nation avoid fragmentation by building a sense of community also played an important role in the drive for national broadcasting. Observers believed new communications technologies could retard the balkanizing forces of immigration and urbanization and reverse some of the disruptive forces of industrialization. In this last sense, radio would provide a technological solution to problems caused by previous technological developments. The large corporations did not simply impose national service on a passive audience of consumers. A number of radio listeners seemed to prefer national service because they wanted to hear broadcasts from distant places, especially from cities that offered different forms of programming. This demand was an important legacy from the early days of radio when "distance fiends" competed to see who could pick up stations located the greatest distance from their receivers.[40]

One of the new techniques for providing national service, interconnection, attempted to use wires or shortwave radio signals to transmit broadcasts originating elsewhere to a station for rebroadcast. AT&T made the first major effort to use telephone wires during the fall of 1922 with broadcasts of sporting events by distant stations. The company had made plans as early as December 1921 to develop a system of radio stations connected by the company's long-distance telephone lines. In 1922, AT&T set up its first station, WEAF, in order to experiment with wired interconnection. During the next three years, WEAF made a number of arrangements with other stations to carry different kinds of program material sent over the company's telephone lines specially adapted for the task. These experiments demonstrated the feasibility of constructing networks of stations providing national coverage. In his opening address at the third radio conference, Hoover called interconnection "the greatest advance in radio since our last conference."[41]

During 1923, RCA in particular began seriously to consider the use of high-power stations to achieve national coverage. The company's vice president, David Sarnoff, predicted that a system of three to six superpower stations simultaneously broadcasting the same programming from specially chosen locations could provide high-quality national service to every home in the country. Although many questions remained about the feasibility of the proposal, Sarnoff thought the plan had a number of advantages, including lowered costs that would result from station consolidation. Broadcasters and government officials loosely used the term *superpower* to refer to stations operating at up to 50,000 watts (generally 5,000 watts and higher). Hoover asked the third radio conference to evaluate a number of requests from stations to operate at these higher levels. Small stations not controlled by the large communications companies felt threatened by both superpower and interconnection and warned of the dangers of monopoly. At the opening of the third radio conference, Hoover tried to reassure these small stations that he would look out for their interests, but the problem of monopoly control became a major source of controversy at the conference.[42]

Despite the concerns about a trend toward economic consolidation that might threaten small stations, the large companies that posed the major threat were by no means united. As early as 1922, a division began to develop within the radio industry between what became known as the Radio Group (mainly RCA, GE, and Westinghouse) and the Telephone Group (AT&T and its subsidiary Western Electric). This development helps explain Sarnoff's campaign to develop superpower as a technological alternative to the phone company's experiments with wired networks. The main reason for the split between the two groups was a difference of opinion about how to interpret the cross-licensing agreements developed after the formation of RCA, especially as they applied to broadcasting. The question of who would control different aspects of broadcasting had never been fully spelled out because industry representatives had drawn up the accords before the new use for radio had become a reality. By defining broadcasting as an extension of radio telephony—in this case transmission to a large audience—AT&T argued that its exclusive control of patents to telephone transmitters also covered commercial radio broadcasting. Stations would need to obtain licenses from the company to operate transmitters. Further, other companies would not be able to use the telephone lines controlled by AT&T for interconnection. RCA and the other members of the radio group sought technological alternatives to interconnection through wires because AT&T refused to allow the other companies to experiment with its telephone lines. They tried telegraph lines, but this alternative was less satisfactory for transmitting the human voice. Sarnoff's promotion of superpower during 1923 may have been timed to influence negotiations with AT&T over the original patent agreements. While RCA experimented with superpower, GE and Westinghouse attempted, with limited success, to develop the use of shortwave radio to interconnect stations. In response to the claims of AT&T, members of the Radio Group contended that broadcasting depended mainly on receivers that they had an exclusive right to manufacture and sell. They believed that a fundamental difference existed between commercial broadcasting and telephony.[43]

In addition to pioneering early efforts to use interconnection for national service, AT&T also set an important precedent in the use of advertising, or toll broadcasting, to defray expenses. Its New York station, WEAF, charged other commercial firms a fee for the use of its facilities to sell products. In contrast, members of the Radio Group had first become involved in broadcasting as a way to sell receivers to the public. Indeed, by 1923 nearly half of the more than five hundred stations operating were associated with manufacturers and electrical firms. But the prohibitive cost of producing programming comparable to WEAF's broadcasts forced these stations to rely on advertising sponsors. RCA formally approved the selling of airtime to advertisers in July 1924. At the third national radio conference, Hoover continued to speak out against direct advertising, but he now accepted as a reality the industry's move toward the use of advertising-sponsored programming, or toll broadcasting. He opposed alternative schemes to tax receivers or impose a charge on listeners and said he believed that public experimentation would produce the best system. According to Hoover, "if radio broadcasting shall be overwhelmed with advertising the radio audience will disappear in disgust."[44]

The debate over interconnection and superpower broadcasting, especially the latter, which dominated the proceedings of the third radio conference, again illustrates the fundamental tension involving the role of technical evaluation and legitimation. Proponents of superpower argued on technocratic grounds that high power was desirable because it would bring the strongest signal to the largest number of people. They assumed that high power could be equated with high-quality programming. Goldsmith and other engineers played an important role promoting these technocratic assumptions at the conference. Goldsmith contended that regulation "must be guided by the idea of the greatest good for the greatest number." The superintendent of radio operations at Westinghouse, C. W. Horn, testified that "the public wants the big features that occur, wherever they occur." Hoover had used the radio conferences to emphasize that different broadcasts should be evaluated based on how well they supported the public interest. According to Hoover, "radio communication is not to be considered as merely a business carried on for private gain, for private advertisement or for entertainment of the curious. It is a public concern impressed with the public trust and to be considered primarily from the standpoint of public interest."[45] Proponents of high power thus set an important precedent by evaluating the public interest standard in instrumental, quantitative terms.

In testimony at the 1924 radio conference, supporters also argued that superpower should be authorized because it represented an inevitable aspect of the progress of science and technology. Sarnoff maintained that "power was the driving force of radio development." According to the RCA vice president, "if increased power is an advancement of the art, it's bound to come whether we oppose it or not." Goldsmith stated explicitly that high-power broadcasting was a "long step forward in the orderly and inevitable evolution of broadcasting in the United States." Sarnoff and Goldsmith both believed that the great progress made in radio research and development in the United States had been possible because of the "American principles of unfettered industry and untrammeled research." The terms *autonomous technology* and *technological determinism* are helpful for making sense of the views expressed by these two men. They believed technology was an autonomous force having an internal logic driven by quantitative concerns such as transmitter power. According to this view, engineers familiar with the inner workings of technological progress should have authority to dictate policy.[46]

In evaluating high-power broadcasting, Sarnoff, Goldsmith, and other participants wanted the conference to draw a sharp boundary between technical and nontechnical considerations. Since they believed broadcasting was fundamentally a technical problem, they thought that engineers capable of evaluating technical questions, such as the amount of interference superpower transmitters would produce, ought to decide whether the government should authorize high-power stations. When the economic or political problem of monopolies was brought up, Sarnoff pushed the conference to take a stand on "whether it is sitting to discuss the question of monopolies, or whether it is sitting as a scientific body to discuss the technical questions." He thought it was clear that the conference was supposed only to evaluate technical problems; the Federal Trade Commission or Congress, not the Department of Commerce, should evaluate monopoly control.[47]

Technical experts, according to proponents, were most qualified to evaluate high-power broadcasting not only because they understood the technical details better than other individuals but also because they were "supposed to be free from any kind of bias." Such was the reason given by the chairman of one of the subcommittees for asking John Dellinger, head of the radio department of the Bureau of Standards, to testify on the "purely technical side" of high-power broadcasting. Other engineers, especially Goldsmith, used the occasion to stress their important role in advising the Department of Commerce about radio matters. Goldsmith—who had been employed by RCA since its establishment, serving as chief broadcast engineer beginning in 1923—argued that because the "engineering matters" of high power were "such that no one discussion could possibly cover the subject," they "must be handled through a continuing committee of some sort." To reassure individuals who might have doubts about the impartiality of engineers who were in many cases employed by companies wanting to set up superpower stations, he emphasized the "special" nature of the Institute of Radio Engineers, the professional association representing radio engineers: "It may be mentioned that the Institute of Radio Engineers is an entirely non-partisan and scientific society, free from any commercial affiliations or connections whatsoever." A GE executive testifying at the radio conference confirmed that one of his employees was attending as a representative of a professional engineering association, not as a company employee.[48] To call the Institute of Radio Engineers nonpartisan seems very simple-minded, but participants at the radio conference assumed that a sharp distinction could be made between the two roles for engineers.

Although Sarnoff believed the conference participants could be divided into "two classes, the non-technical men, who have been against raising power, and the technical men, who have been for it," at least one important engineer, Cyril Jansky, joined opponents of superpower who urged the conference to take into account not only the engineering considerations but also the "economic, political, and social aspects." Jansky and other opponents wanted the Department of Commerce to evaluate the effect of superpower broadcasts from larger cities on local, low-power stations. Such groups as the Citizens Radio Committee, which claimed to represent "the great army of listeners-in, retail radio dealers, and independent broadcasters," warned that superpower would not only drown out local stations but also monopolize radio channels. The citizens groups, as well as a collection of radio stations owned by newspapers, asserted that this would lead to a restriction of program choice and the destruction of individual rights. Although one argument for superpower emphasized that it would benefit listeners in rural areas, at least one citizen from rural Tennessee contended that "it is the city listener-in that is calling for more power—not the ruralist."[49]

Opponents of superpower specifically questioned the technocratic views of men such as Sarnoff and Goldsmith. Instead of assuming that all technological development could be equated with general social progress, critics of high power wanted the conference to ask the question: Progress for whom? The strongest attack came from C. E. Erbstein, a small-town lawyer who was participating as a representative of a radio district in Illinois. Erbstein alleged that the conference was dominated by "the four horsemen of progress and advancement in the science of radio"—that is, RCA, Westinghouse, AT&T, and GE. He championed the cause of the average citizen and questioned whether the participants were really interested in supporting the public interest: "Do they wish to foster all this advancement in science and radio on the public for the good of the public, or for the good of the corporations that they represent?" Erbstein and others doubted the engineers' claims that they could put aside their employers' interests and give unbiased evaluations based on their engineering consciences. "Who are the people here?" he asked. "Who are for high power? What are their connections? Let's call the roll. . . . Let's take them one by one. . . . And you either find a connection with one of the 'four horsemen' of radio, or you will find one of them who has already purchased a five kilowatt set."[50]

Unbiased evaluation was also the goal of high-power opponents, but they proposed an alternative to reliance on engineers. They used public surveys to support their views. A participant from Boston claimed that his "position on the matter of superpower [was] taken from the viewpoint of the listener-in entirely, without any personal motives dictating my position." Opponents of high power viewed public experimentation as preferable to the theoretical and idealized predictions of engineers presenting "scientific papers on the subject." They claimed that engineers made unrealistic judgments about interference based on "superior equipment . . . able to tune out superpower stations," which the average listener did not own. Opponents also argued that broadcasters could still develop national systems by interconnecting local stations instead of using superpower.[51]

Under pressure from opponents of superpower, Sarnoff withdrew his request that the Department of Commerce allow unlimited power. He agreed with the position favored by Hoover that further testing should be done before the department granted final authorization. Sarnoff told the conference that RCA was willing to absorb the costs, but he also emphasized the important role of technical experts in evaluating the public experiment: "If, after a reasonable opportunity to test that station—and that can be defined by experts in the art—that station proves to be undesirable from the public standpoint, we will shut it down and pocket the loss." According to Sarnoff, critics of superpower were inconsistent in supporting interconnection but opposing superpower because of worries about monopoly control. He pointed out that a company such as AT&T, one that controlled the wires needed for interconnection, could also use the technique to gain a monopoly. The conference's final recommendation favored further experimentation with both high power and interconnection. In general, the conference raised issues only after receiving Hoover's approval; the recommendations thus served to legitimate policies the Department of Commerce had already developed.[52]

New Challenges: Crisis and Legislative Action

Discussions at the third radio conference underscored the major changes that had occurred in the industry during 1924. The Department of Commerce continued to receive requests from prospective broadcasters to use the already overcrowded broadcast band, which the third conference recommended should be extended to include frequencies from 550 to 1,500 kHz. Superpower and interconnection threatened to create new sources of interference. The acceptance of advertising and its success made broadcasting profitable and even more popular. New types of broadcasts helped widen the public interest. Hoover argued that for many Americans, especially in rural areas, radio was "rapidly becoming a necessity." After receiving support for his policies from conference participants, Hoover tried again to convince Congress to pass legislation. Instead of proposing a sweeping new law, he asked Congressman White in December to introduce in Congress a "very short bill clarifying the powers of this department as to radio regulation." The new legislation treated the regulation of broadcasting as a technical problem involving "wave lengths, power, apparatus and time of operation." The proposal "was narrowed down to this field," according to Hoover, "leaving the bigger issues of regulation until we have enlarged knowledge of the art and of the problems with which we are now confronted."[53]

Hoover sought immediate authority because new stations were quickly using up available frequencies in the broadcast band. His department could have attempted to judge based on quality of programming that some stations were more worthy than others for licenses. But Hoover argued that "any attempt to give preference among stations in the allotment of wave lengths on the basis of quality of programs raises the question of censorship, the implications of which I cannot at present accept."[54] By narrowing the problem to the technical issue of deciding how to develop an efficient system of allocating frequencies to different stations, Hoover sought to avoid controversial issues such as censorship and monopoly. In this scheme, engineers would play the major role by deciding how to develop a rational allocation system that would allow the maximum number of stations to broadcast in a limited band of frequencies.

Hoover was unsuccessful in his attempt to gain support for the limited legislation he had suggested. Unlike earlier legislative proposals that had died in Congress mainly because of lack of interest and limited support, this effort met with strong public opposition from a few influential individuals. By 1925, the growing competition between different broadcasters was helping to undermine the consensus Hoover had forged. Eugene F. McDonald, the president of the newly founded trade association the National Association of Broadcasters, warned that Hoover's bill would "vest any Secretary of Commerce with Napoleonic powers." He proposed as an alternative that Congress create an independent commission in charge of all communications. McDonald and other critics were not hostile toward Hoover, but they worried about what other administrators in the future might do if given such expansive power.[55]

Despite the legislative defeat and the attack on his authority, Hoover worked toward implementing a new allocation system. Following the third radio conference, he did away with the class C category, moving the stations to the class A frequencies. After experiments with different separations between frequencies used by broadcasters, the Department of Commerce concluded that it needed to create a ten-kilocycle separation between the channels used by broadcast stations in order to avoid interference. This decision resulted in eighty-six available channels in the broadcast band; the department reserved thirty-nine channels for class A stations and forty-seven channels for class B stations. Because class A stations broadcast for only part of the day at low power reaching distances of only twenty-five to fifty miles, interference was generally not a major problem. Class B stations were a different matter. Since less than one frequency was available for every two of these stations, the department allowed most to broadcast only for part of the day. Further, the department divided the country into five zones and assigned approximately ten of the frequencies available for class B stations to each zone. Beginning in 1925, Hoover allowed some class B stations to increase power by increments of 500 watts toward a goal of 5,000 watts. Officials believed these 5,000-watt stations would have a broadcast radius of several hundred miles. Because he came to define the public interest in terms of national service, Hoover encouraged experimentation with high power and interconnection. According to Hoover, "the local broadcasting station must make available to the audience the greatest music, entertainment and enlightenment the nation and the world affords." In order to determine the implications of high power and interconnected stations on the allocation system, Hoover asked for additional funds from Congress to undertake detailed studies of the service areas of different stations and any interference that might occur. "Such an investigation," he believed, "may disclose possibilities of a better basis of wave length distribution."[56]

By spring 1925, Hoover recognized that a new rational allocation system would not necessarily be able to solve the problems of the industry, especially because the Department of Commerce had on file 425 applications for additional licenses. If all requests were granted, 988 stations would saturate the broadcast band. Faced with these new difficulties, Hoover warned that all applicants should not assume they could continue to receive licenses.[57] Since he did not have clear legal authority to deny licenses to qualified applicants, Hoover followed his tried-and-true practice of mobilizing radio users to support his policies. Toward this end, he held a fourth radio conference in November 1925. More than 450 individuals attended; this compared with fewer than 25 at the first gathering.

The fourth conference evaluated a number of problems, but the key issue was the proposal to limit the number of stations by denying licenses. In his opening address, Hoover drew an analogy to the national highway system to get his point across: "We must face the actualities frankly. We can no longer deal on the basis that there is room for everybody on the radio highways. There are more vehicles on the roads than can get by, and if they continue to jam in all will be stopped." Hoover acknowledged that his policy of supporting high-power broadcasting had helped create a need for restricting licenses. "Higher power has greatly strengthened the service to listeners," he said, "but it has aggravated the problem of providing lanes through the traffic." He also admitted that more stations could operate without creating new interference if the department decreased the amount of time it allowed each station to broadcast, but he believed this would result in "much degenerated service. . . . A half dozen good stations in any community operating full time will give as much service in quantity and a far better service in quality than eighteen, each on one-third time." Another partial solution might have been to broaden the broadcast band, but this would have meant taking over frequencies previously assigned to amateurs. Hoover also rejected this option; he needed the amateurs' support to push new legislation through Congress. The public viewed amateurs as young experimenters who had played an important role in the early development of wireless. Citizens might view an attack on amateurs as an attack on the "American boy." Despite these examples of the role of specific policy decisions in helping to create the new situation, Secretary Hoover mainly justified his judgment to restrict licenses based on "a simple physical fact that we have no more channels." It was a technical necessity; "not a question of what we would like to do but what we must do."[58]

The conference delegates unanimously supported Hoover's decision to curtail station licenses, but the crucial question remained—what criteria should the government use to decide that one station deserved a license more than another? The delegates followed Hoover's advice that radio stations had to demonstrate they were committed to the public interest in order to justify receiving a license. A broadcaster "must perform the service which he had promised," Hoover warned, "or his life as a broadcaster will end." In response to critics who thought he was advocating censorship or placing a limitation on freedom of speech, Hoover pointed out that "there are two parties to freedom of the air. . . . There is the speech maker and the listener." The secretary sided with freedom for the listener: "We do not get much freedom of speech if 50 people speak at the same place at the same time, nor is there any freedom in a right to come into my sitting room to make a speech whether I like it or not." The standard of serving the public interest had its origins in public-utility regulation, but Hoover and the conference delegates emphasized that broadcasting involved different considerations. Most notably, they thought government regulation of rates and other economic factors internal to business operations was appropriate for public utilities but not for broadcasting.[59]

Although the conference delegates agreed with Hoover that broadcasting needed to serve the public interest, they disagreed about how to identify this standard. The tendency of policy makers to follow technocratic views was fundamentally important in the debates about evaluating the public-interest standard. On the one hand, for example, Hoover assumed that the technical or instrumental advantages of high-power broadcasting were sufficient to justify its continuance. In his opening address to the conference, he emphasized that the experiments with "power increase has meant a general rise in broadcasting efficiency." Rather than analyze the social and economic implications of the new policy, especially its effect on smaller stations, Hoover's discussion focused on such advantages as increased signal strength, clearer signals, and the overcoming of static.[60]

On the other hand, Hoover also seemed to acknowledge the existence of two ways to view the problem of regulating broadcasting in the public interest. He drew a distinct boundary between the problem of "traffic control" and "the determination of who shall use the traffic channels and under what conditions." The first issue involved technical questions of administrative rule making aimed at reducing interference, including deciding how to allocate wavelengths and control power. Technical experts would play a key role evaluating these problems. The second issue involved "semijudicial" considerations that "should not devolve entirely upon any single official." According to Hoover, this side of radio regulation demanded that "each local community should have a large voice." He encouraged the conference delegates to recommend the establishment of regional committees that would help decide who should use the frequencies that the Department of Commerce assigned to their regions. Although Hoover and the conference delegates generally argued that the industry should evaluate such issues as the use of direct and indirect advertising and the relative value of different types of programming, the proposed regional committees might, presumably, help evaluate such controversial issues. Some participants at the radio conference, including Jansky, also tried to convince the delegates to recommend that particular groups, especially educational broadcasters connected with universities and colleges, had "a right to special consideration when it comes to the question of time and wave length." The educational broadcasters thought regulation of such issues as high-power broadcasting or wavelength allocation should not assume a sharp boundary between technical and nontechnical considerations. In their view, the two realms were necessarily interrelated.[61]

Although some of the separate committees at the conference supported Hoover's proposal for regional policy groups, influential delegates representing large manufacturers and broadcasters succeeded in scuttling the idea. Company representatives contended that radio broadcasting was inherently national in scope. They believed the establishment of regional committees would be incompatible with their effort to promote national broadcasting. "There is no such thing as a region in broadcasting," argued Owen Young, chairman of the board of RCA. He complained to Hoover that "inasmuch as I have my heart set, as you know, on a national broadcasting program through several stations advantageously located, in addition to that provided by any superpower station, I should be very sorry to have regional committees dealing with any questions affecting broadcasting." Opponents thought regional committees would only inject local controversies and disputes into the industry. They favored a national committee advising the Department of Commerce on technical issues because it would be "nonpartisan" or "free from politics."[62]

Hoover's effort to distribute decision making to different regions went against the general trend in the industry toward consolidation; his support of higher power and interconnection had, however, promoted the consolidation development. During summer 1925, Hoover allowed tests for the first time of a 50,000-watt station, operated by GE. At the time of the third radio conference, only two stations had been outfitted to broadcast using more than 500 watts; by the time of the fourth conference, thirty-two were equipped to operate at 1,000 watts, twenty-five at 5,000 watts, and two at even higher levels. Supporters continued to argue that high power was a "symbol of the progress of the science" of radio, and they warned that it would be "the greatest kind of mistake to meddle with a pioneer art or to attempt to stop scientific and technical development." Proponents also stressed the importance of high power and interconnection for rural listeners, who otherwise would not have access to high-quality programs from major cities. They tended to ignore the fact that educational stations at agricultural universities that were specifically serving rural areas had strong reservations about high power and similar efforts toward economic concentration.[63]

Industry consolidation was also driven by the conflict between the Radio Group and the Telephone Group, which involved different interpretations of their cross-licensing arrangements. During 1925, the two combinations worked out final arrangements that included a decision by the Telephone Group to withdraw from broadcasting in exchange for exclusive rights to public-service telephony. AT&T officials feared that growing antimonopoly sentiment against the radio trust might also turn on their company if they continued to pursue broadcasting. RCA, GE, and Westinghouse received exclusive rights to radio telegraphy and the broadcasting of entertainment; the companies also maintained control of the manufacture and sale to the public of radio tubes. In 1926, AT&T sold its New York station WEAF to RCA for $1 million; in return it retained an exclusive right to provide the use of telephone lines for a fee to broadcasters. Using AT&T wires to distribute programming to different stations, RCA established, in 1926, the first major network, the National Broadcasting Company (NBC). RCA owned 50 percent of the new company, GE 30 percent, and Westinghouse 20 percent. RCA continued to serve as a sales agent for apparatus manufactured by the other companies, but now it also owned a network of broadcast stations. With the added expense of leasing telephone lines, RCA needed advertising revenue more than ever, and with the ability to pursue wired interconnection, the company no longer had the same incentive for developing superpower national broadcasting. Even if RCA had not acquired network capability, by the late 1920s, the feasibility of Sarnoff's vision of a handful of stations blanketing the entire country with interference-free broadcasts seemed doubtful.[64]

Following the fourth radio conference, Hoover felt he had the necessary support for new legislation to replace the 1912 Radio Act. In December, Congressman White and Senator Clarence C. Dill, Democrat from Washington, introduced nearly identical bills in Congress. The proposed legislation incorporated the recommendations of the radio conference and reflected White's earlier efforts to give the Department of Commerce wide authority to regulate broadcasting, including the power to limit licenses. The bills also rejected the idea of regional committees in favor of a national radio commission that would advise the secretary of commerce on difficult issues and listen to appeals from disaffected broadcasters. Other provisions did not allow broadcasters to claim ownership of wavelengths; their applications for licenses would be judged based on their commitment to the public interest.[65]

The White bill passed the House in March 1926, but the public debates indicated that Hoover did not have support from all the industry. Despite the provision for advisory committees as a check on arbitrary authority, McDonald and other influential individuals continued to warn that Hoover might become the "czar" of radio. They proposed as an alternative the creation of an independent commission, modeled on the Interstate Commerce Commission, to regulate all communications. Other critics included educational broadcasters allied with the Department of Agriculture, who threatened to withhold support unless the new law gave them preferential treatment. Similarly, owners of small independent stations claimed the legislation was written to help members of the "radio trust" consolidate their position in the industry by forcing the independents off the air. Congressman Ewin L. Davis of Tennessee also complained that the sections of the proposed legislation addressing the issues of monopoly control and restraint of trade were too weak. He favored reintroducing language from earlier bills that would give the secretary of commerce authority to deny licenses to companies or individuals based on his own judgment that they were attempting to gain a monopoly. Both the White bill passed by the House and a Dill bill being considered by the Senate seemed to allow the secretary to act only after a federal court had made a ruling against an applicant.[66]

In part due to a preoccupation with issues that seemed more immediately pressing, the Senate failed to act on the proposed legislation early that year. But in April, a new development forced a reevaluation of the industry and its relationship to the government. In January 1926, the maverick broadcaster Eugene McDonald, president of Zenith Radio, challenged Hoover's authority by directing his Chicago station WJAZ to begin broadcasting on an unauthorized frequency that the Department of Commerce had set aside for a Canadian station. The department took Zenith to court, charging it with violating the 1912 Radio Act, but on April 26 a U.S. district court in Illinois ruled in Zenith's favor. Significantly, Judge James H. Wilkerson argued that the Radio Act did not give Hoover authority to assign frequencies, power levels, and hours of operation to stations or to deny licenses to applicants. This decision effectively led to the dismantlement of the entire administrative structure Hoover had spent years constructing.[67]

After the Zenith ruling, Hoover argued that the possibility that station owners might create a chaotic situation by broadcasting wherever and whenever they pleased demonstrated the need for immediate action by the Senate on radio legislation. But by May, Senator Dill and his supporters had decided to revise their proposed bill, incorporating McDonald's recommendation that an independent agency, rather than the Department of Commerce, oversee broadcasting. Hoover lost support because senators from both parties worried about the implications for the upcoming presidential election if Congress gave him control of communications. Both sides predicted an important role for radio in the campaign, which they expected Hoover to enter. Democrats worried that a Republican secretary of commerce might deny their party access to the airwaves. "Hoover, if he runs for president," wrote one newspaper editor, "will have as . . . supporters every broadcasting station which has a permit and which hopes to get a renewal of its license." Key Republican senators, including the majority leader and chairman of the Committee on Interstate Commerce (who opposed Hoover's nomination to the Republican ticket in the 1928 election and who viewed him, in the event that Coolidge did not run for reelection, as a frontrunner) were also very willing to support any legislation that might weaken the secretary's political position. President Coolidge weighed in against the Dill bill, on the grounds that he opposed setting up any additional federal agencies. Another opponent, the important engineer and inventor Michael Pupin, expressed a technocratic sentiment when he argued that "the Senate is wrong . . . when it proposed to solve a complicated scientific problem in its own way without any knowledge of the science . . . my message to the Senate is—Hands Off!" Despite these efforts to defeat the new proposal, the Senate passed the Dill bill on July 2. Since the legislative session ended the next day, senators did not have time to work on a compromise with the White bill in the House, which was nearly identical except that it gave authority to the secretary instead of to an independent commission.[68]

Pupin's comments underscored an important theme that became evident in the debate over the two measures. One of the major reasons supporters of the Dill bill wanted an independent commission was to place major decisions in the hands of technical experts to avoid partisan interference. This had been the motivation for the creation of other independent regulatory commissions in the United States, including the Interstate Commerce Commission and the Federal Power Commission. Charles Francis Adams, the brother of the novelist Henry Adams and one of the original promoters of the idea of independent commissions during the late nineteenth century, argued that "commissions might scientifically study and disclose to an astonished community the shallows, the eddies, and the currents of business." The movement to form regulatory commissions was an aspect of the larger effort during the late nineteenth and early twentieth centuries to respond to instabilities resulting from transformations connected to the rise of a national economy based on corporate capitalism. Both business and government leaders were convinced that unrestrained free enterprise did not necessarily result in a rational economic system. Progressive Era government agencies and commissions attempted to construct rules and standards to rationalize business behavior. In many cases, business executives supported, even encouraged, efforts to regulate industry and create stability. An independent, expert commission seemed especially appropriate for radio; unlike the power or transportation industries, radio seemed to have an even closer connection to scientific developments. Other commissions needed technical experts, including engineers, primarily to determine economic problems such as ideal rates and allowable profits; a radio commission would need engineers and scientists to apply knowledge of the physics of radio-wave propagation and circuit design to such problems as station interference and spectrum allocation.[69]

Ironically, Senator Dill disagreed with supporters of his legislation who thought commissioners on the proposed radio commission "should be experts." Rather, he believed the best candidates would be "men who have an understanding of the public needs, men of vision and great ability." According to Dill, speaking that summer, "I do not think it would be wise to have a commission made up of technical experts, because technical experts would not take the big view and the broad view and have the vision which I think the members of this commission ought to have." He did not presume that engineers could "lift themselves above the technicalities." He also seemed to be skeptical of engineers' claims that they could make decisions based on the public interest rather than the interests of their employers. Experts would be important, according to Dill, but the commission should use them as advisers providing technical evaluation that the commissioners could interpret in order to take into account "the future development of the radio art for the social and economic good of our people." In fact, Dill did not think the commissioners needed to know anything about "science as such, or the technical side of radio," because the problems they would have to address fundamentally involved "social and economic" considerations.[70]

Congress did not reconvene until December 1926. During the intervening months, the lack of legal authority for the regulation of radio broadcasting resulted in near chaos in the spectrum. The number of stations operating in the United States increased from 528 in December 1925 to 719 a year later. Sixty-two stations switched frequencies and sixty-three increased power. Hoover had instructed the department to obey the decision in the Zenith case and not interfere in station operations after he had received an opinion from the attorney general, in July, that confirmed the court's decision.[71] It seems clear that he hoped this action would put pressure on Congress to pass necessary legislation.

By December, Congress was ready to take action to prevent further chaos. Radio broadcasting had become an important public issue as citizens flooded their representatives with letters complaining of interference. Existing evidence also indicates that key members of Congress worried that if they did not act, another court decision, known as the *Oak Leaves* case, that had been handed down after the Zenith ruling might set an important precedent for the legal establishment of a market system based on private-property rights in the spectrum. Many members of Congress still believed strongly that the spectrum was a special resource that the public should continue to own because of its unique role in conveying information and molding opinion. Aitken later pointed out that they wanted to protect the spectrum from commercial exploitation because "as more than one senator expressed it, the spectrum was the last remaining public domain, and it was scarce in a sense in which public land never had been." With these concerns, House and Senate conferees quickly agreed to a compromise bill on December 21. The new measure backed the establishment of an independent agency, which became known as the Federal Radio Commission, but Congress authorized funding only for one year. If Congress had not taken further action at the end of this period, authority would have reverted back to the Department of Commerce. In February 1927, after the House of Representatives and the Senate voted in favor of the new measure, President Coolidge signed the new Radio Act into law.[72]

The Radio Act of 1927 gave the Federal Radio Commission extensive authority to regulate radio broadcasting. Specifically, the law gave the commission authority to limit interference and keep order in the airwaves by assigning frequencies, power levels, and times of operation to stations. The commission was also expected to classify stations and fairly distribute licenses to all regions of the country. Unlike the 1912 act, which assumed that all citizens had a right to a license, the 1927 act emphasized that broadcasting was a privilege given to individuals based on their commitment to "public interest, convenience, and necessity." The act stated that all stations had to reapply for licenses and demonstrate to the commission they were serving the public interest. The commission would issue licenses for only three years, and if a station violated any provision of the act, the commission could revoke its license. Section 29 reassured broadcasters that "nothing in this act shall be understood or construed to give the licensing authority the power of censorship." Section 13 took a strong stand against monopolies and unfair trade practices, but seemed to maintain that the commission could deny a license only after a federal court had determined that an owner had violated the law. Finally, to protect broadcasters, section 16 allowed them to appeal commission decisions to the Court of Appeals of the District of Columbia, which had authority to "alter or revise" the commission's rulings.[73]

Chapter 2 analyzes the crucial decisions of the Federal Radio Commission, which effectively authorized what became popularly known in commercial broadcasting as the "American System." However, the key themes introduced in this chapter, including the tension between technocratic and nontechnocratic views about regulating broadcasting, are especially important for understanding this development. In one sense, Hoover tried to manage this tension by having it both ways. Especially at the third and fourth radio conferences, for instance, he both emphasized the key role of engineers and tried to include as many different groups with an interest in radio as possible. But the engineers' involvement cannot be underestimated. Writing in 1928, Goldsmith recalled with pride how engineers' recommendations had been of central importance at the radio conferences. Representatives from the Institute of Radio Engineers, he related, had "participated in the formulation of the recommendations which [were] informally adopted as guides by the Department of Commerce in the administration of the law [and] were markedly instrumental in the rapid development of the radio art."[74] Engineers served a critical function mediating and harmonizing the close relationship between government and private industry.

The tension between the two outlooks on radio policy is evident also in the problem regulators faced of deciding what kind of criteria to use in policy making. On the one hand, officials tried to reduce radio regulation to the technical problem of combating interference. On the other hand, Hoover and others connected with the Department of Commerce also on occasion acknowledged the relevance of social, economic, and political factors. But they tended to draw a sharp distinction by not examining possible interrelationships.

The next chapter explores more fully how an emphasis on technical evaluation helped legitimate complex decisions establishing a national system of commercial broadcasting. Technocratic values influencing the Federal Radio Commission helped finesse traditional opposition to a system supported by commercial advertising. But a different aspect of commercialization, the establishment of de facto property rights in the spectrum, had already become well established by the late 1920s. The logic of technical necessity again played a crucial role in helping to circumvent traditional opposition to the treatment of the spectrum as a commodity. The Department of Commerce set a precedent for future regulation when it routinely allowed stations being bought and sold to include the transfer of its broadcast license. The cost thus reflected not only the value of equipment and other material possessions but also the frequency assignment, licensed power levels, and other broadcast authorizations the government had previously granted to the original station. When in 1926 RCA bought WEAF, AT&T's New York radio station, only one-fifth of the $1 million paid went toward the building and broadcast equipment. The large price mainly represented compensation for the valuable spectrum rights of a clear-channel station.[75]

This de facto market for spectrum rights has developed partly because the standard rationale for broadcast regulation first developed by Hoover and written into the 1927 Radio Act did not explicitly emphasize the political conviction, expressed by many members of Congress during the 1920s, that the broadcast spectrum was a unique resource that private individuals should not market commercially. The public rationale emphasized that government regulation of broadcasting was a technical necessity demanded by the physical fact of "spectrum scarcity." Policy makers justified government intervention based on the shortage of channels available in the broadcast spectrum; all stations could not broadcast without causing severe interference. Later Supreme Courts have upheld the legality of this reasoning. The logic of technical necessity thus helped finesse an apparent contradiction between political rhetoric and economic reality. Although Congress enacted legislation in 1927 to avoid commercial exploitation of the spectrum, the main result has been to establish de facto rights for large commercial broadcasters using sophisticated and expensive equipment.[76]

2)))

Radio Engineers, the Federal Radio Commission, and the Social Shaping of Broadcast Technology

"Creating Radio Paradise," 1927–1934

As consultants, as moulders of the policies of great industrial organizations, as expert witnesses and legislative advisers, the radio engineers, with their group consciousness, can be of service as great and important as with their individual ingenuity in the laboratories.

Assistant general counsel, Federal Radio Commission, June 1930

During the late 1920s and early 1930s, the tension involving the tendency to follow technocratic values in order to rationalize decision making was resolved, and the way in which this happened had an important influence on the development of radio broadcasting in the United States. Specifically, the dominant technocratic decisions of the commission helped support the growth of a technological system of commercialized, corporate-controlled network radio. The most important technocratic policies of the Federal Radio Commission (FRC) involved the development of a new rationalized allocation system for radio broadcasting, and the final section of this chapter relates my historical discussion of this topic to tensions in engineering professionalism. An examination of how these tensions were resolved in the work of the radio engineers advising the FRC helps illuminate the political, economic, and social values embedded in the technocratic policies of the commission.[1]

The Radio Law of 1927 established the commission as an independent agency to regulate the industry. Congress expected the commission to use the objective methods of technical decision making to help keep its work free from political influence. The commission sought to rationalize radio broadcasting by imposing order on a chaotic situation. Stations that could not demonstrate they were serving the public interest faced revocation of their licenses. By helping to reinforce particular trends already under way in the growth of the industry, the FRC was instrumental in establishing the political, economic, administrative, and social characteristics of the American system of broadcasting. The 1934 Communications Act, which established the Federal Communications Commission (FCC), sought to centralize the government's administration of communications in one federal agency; as far as the regulation of radio broadcasting was concerned, the 1934 law essentially institutionalized the policies of the radio commission.[2]

The central dilemma of government regulation of radio that both Hoover and the radio commission tried to solve was that the number of broadcast channels was clearly limited. To complicate matters, this number was never an absolute quantity but tended to increase over time, with new developments in radio instrumentation and technique. During 1928, approximately ninety-six channels existed in the broadcast frequency band (that is, between 550 and 1,500 kilocycles). Different factors placed limitations on the number of stations that could be on the air at any one time. One problem was that signals actually extended beyond the specific frequency of transmission. Telegraphic signals took up a few tenths of a kilocycle; the broadcasting of music or conversation took up about ten kilocycles.

The frequency difference broadcasters needed to maintain between any two stations depended on a number of additional factors: the power of each station, geographical separation of stations, the selectivity of receivers, and the location of receivers in relation to interfering stations. Two different stations might share the same frequency if they were far enough apart, but a phenomenon known as heterodyne interference complicated matters greatly. Most transmitters in the 1920s were unable to prevent a certain amount of drift to nearby frequencies. When this happened, the slight frequency difference between two stations that were supposed to be operating on the same frequency would create a "beat" frequency: at night, a receiver would normally be able to pick up a good signal from a 5,000-watt station about one hundred miles away, but it would hear the heterodyne interference from two 5,000-watt stations assigned to the same frequency upwards of three thousand miles away. Regulatory issues were different for daytime broadcasts. Thus, the engineering considerations that experts needed to take into account were by no means simple and clear-cut. According to the chief radio engineer for the commission:

> All of the engineering work involved in federal radio regulation ha[s] the peculiar difficulty that the facts dealt with are extremely complex. They are indeed rapidly shifting. Not only must allowance be constantly made for the flux of changes inherent in a rapidly developing art, but radio waves themselves exhibit extraordinary vagaries. Orderly radio regulation must proceed on a consideration of the distances at which the waves are received. But distances vary enormously between day and night, from season to season, even from night to night, and are different over different kinds of terrain.[3]

Although radio engineers often portrayed themselves as technical experts providing decisive solutions, the complexities and contingencies of the problems in radio regulation precluded such definitive results. Engineers made recommendations that involved something more than rule-governed technical judgment.[4]

Radio Engineers and the FRC

Radio engineers played an increasingly important role in the policy work of the commission as its mission became more permanent, its responsibilities more complex, and its decisions more technical. In 1932, the commission held 177 formal meetings and received more than 40,000 applications (during 1929, by contrast, the commission had received 6,927 applications). Not all of these applications were for radio broadcasting; it should be kept in mind that the commission was responsible for regulating all uses of radio. When Congress first established the radio commission, a total of 18,119 different transmitting stations existed in the United States; only 733 of these operated as broadcasters. Nevertheless, the commission considered the regulation of radio broadcasting its "principal task."[5]

The most important government radio engineer who served as a technical consultant to both the Department of Commerce and the FRC was John Howard Dellinger. Dellinger first joined the National Bureau of Standards in 1907, at the age of twenty-one; six years later he received a Ph.D. in physics from Princeton University. As the chief of the Radio Section of the Bureau of Standards (beginning in 1921) and president of the Institute of Radio Engineers (in 1925), Dellinger actively participated in Hoover's radio conferences. He drafted the reports for each of the four conferences and played an especially important role in devising policy for frequency allocation. During the 1920s, Dellinger also served as a delegate to international radio conferences, including the 1921 Interallied Technical Conference on Radio Communications in Paris. In this and in a number of other instances, he was responsible for explaining the U.S. government's decisions to the American radio engineering community and for soliciting their advice. As a public engineer, Dellinger wrote popular articles and presented talks over the radio explaining the work of the Bureau of Standards and speculating on such subjects as the future of radio and the social role of engineers. These different experiences served as a foundation for the key role he played as technical adviser to the radio commission.[6]

In February 1928, the director of the Bureau of Standards officially authorized Dellinger to work "part time" for the commission. But this action only formalized a consulting arrangement that had existed since the establishment of the FRC during the previous year. Dellinger met with the commissioners a number of times during 1927 and served as a member of a special committee of the American Engineering Council that had originally been organized to provide the Department of Commerce with technical advice. The commissioners made Dellinger's unofficial role as chief engineer for the FRC official in August 1928, when they placed him in charge of the newly created Engineering Division. As chief engineer until 1929, Dellinger helped forge a strong position for the Engineering Division within the FRC and institutionalized a close connection to other technical divisions of the government, especially the Radio Section of the Bureau of Standards.[7]

Dellinger was not only an important technical representative of the government, he was also an influential member of the radio engineering community. He met with members of the radio commission during the winter and spring of 1928 and helped convince them of the necessity of taking advantage of the technical knowledge and experience of the engineering community. He also established mechanisms to ensure that this expertise would guide the policies of the commission. In April, the FRC secretary argued that "the best engineering talent obtainable to represent the Government in dealing with the high paid engineers of the large broadcasting companies and other experts is of the greatest importance if the wireless communications is to be placed on a fair and equitable basis."[8]

The main professional organization representing the radio engineers in the United States was the Institute of Radio Engineers. The commission first accepted an offer from the institute to provide technical advice in July 1927. The advisory role of the institute was institutionalized nine months later when it established a broadcast allocation committee to help the commission with its major responsibility—a rearrangement of the allocation of radio stations in the broadcast spectrum. Dellinger was one of the four original members of the committee. These four engineers were especially influential participants in a series of meetings of technical experts in the spring and summer of 1928 that resulted in a plan for reallocating broadcast stations.[9] This plan, which became known as General Order 40, had a major impact on radio broadcasting in the United States.

The status of the broadcast committee was formalized in fall 1928 and its membership expanded to at least six members. Alfred Goldsmith, the president of the Institute of Radio Engineers in 1928, described the committee as a group of "qualified men who would from time to time consider a great number of questions on which the Government and the public desired impartial technical advice of the highest quality." The main responsibility of the institute's broadcast committee after the announcement of General Order 40 in August 1928 was to advise the commission about such problems as "the location of high-power stations with respect to populous areas," the "permissible deviation of carrier frequency from licensed frequency," "the fidelity of transmission," and the "service area of stations of various powers." The committee solicited advice from experts in different fields and submitted preliminary reports to the twelve-member institute board of direction, which drafted the final reports submitted to the commission. All together, from "ten to a hundred or more engineers" contributed to each of the reports.[10] The final reports presented the technical facts and made recommendations meant to guide the commission in formulating regulations.

The commission also received technical advice from three trade organizations: the National Association of Broadcasters, the Radio Manufacturers Association, and the National Electrical Manufacturers Association. But on the recommendation of the president of the Institute of Radio Engineers, Dellinger decided not to ask these three organizations to participate in committees providing technical advice to the commission. Goldsmith argued that the institute was "in a unique position" to legitimate policy decisions by furnishing "authoritative engineering data through its broadcast committee."[11]

At different times, the commission employed the four members of the original institute broadcast committee directly as consultants. Laurens Whittemore played a notable role in the spring of 1928 while an employee of the American Telephone and Telegraph Company (AT&T), which he had joined in 1925 after working for seven years with the Radio Section of the Bureau of Standards. Although Dellinger chaired the engineering conferences sponsored by the radio commission in spring 1928, Whittemore helped recommend which engineers to invite. Whittemore was also instrumental in the formulation of General Order 40.[12]

Robert H. Marriott chaired the institute broadcast committee during the spring of 1928 and the commission appointed him as a consultant in November of that year. He was a well-respected consulting engineer who had served as the first president of the institute. While an employee of the commission, his most important responsibility was to coordinate the work of the broadcast committee to satisfy the needs of the commission. During 1928 and 1929 he conducted studies of special broadcast problems and translated the technical advice of the broadcast committee into a book of recommendations. He also represented the commission at the first meeting, in 1929, of the committee on radio law of the American Bar Association. Dellinger, Marriott, Whittemore, and other principal institute engineers advising the radio commission had been actively involved in radio regulation since at least the early 1920s.[13] As their involvement expanded with the commission, they helped make policy that affected the growth and character of the industry.

Technocratic versus Nontechnocratic Policies

An analysis of the use of radio engineers as technical consultants for the commission reveals the important tension (already discussed) between technocratic and nontechnocratic views. On the one hand, the commission had the option of trying to regulate the social, economic, and political aspects of broadcasting. Decisions about who should receive licenses to broadcast could, for example, be based on such qualitative issues as the character of broadcasts, the use of advertising, the value of large-scale networks, and the educational benefits of programming. On the other hand, the commission could try to reduce regulation to its technical aspects by defining it in purely instrumental terms. Engineers would play a central role by constructing the most efficient system of allocation of frequencies and power.

During the first year of its existence, the FRC recognized that it had a responsibility to outline "a few general principles" that would guide its regulation of radio broadcasting based on the only major standard that the 1927 Radio Act gave the commission as a guide for making decisions.

Congress directed the commission to use the test of "public interest, convenience, or necessity" to evaluate applications. This standard was meant to place limitations on the power of the commission. However, because the statute never attempted to define public interest, convenience, or necessity, the FRC was generally free to make its own decisions.

The commission generally interpreted a number of the principles it used as dealing with the technical aspects of broadcasting. The commission stressed that public interest would be served by "such action on the part of the commission as will bring about the best possible broadcasting reception conditions throughout the United States." The goal was to reduce interference, using, by implication, the most efficient system that would benefit the most people. To avoid interference, the FRC expected stations to follow high technical standards. Transmitters, for example, needed to be of sufficient quality to minimize drift to other frequencies. Officials also needed to regulate the location of stations to avoid interference. Finally, the commission believed that there should be a "fair distribution of different types of service," including a class of high-power stations that would provide a "high order of service over as large a territory as possible."[14]

In addition to these more exclusively technical, quantitative considerations, the radio commission outlined qualitative principles that would guide its interpretation of the public-interest standard. The commission emphasized that it generally did not condone broadcast stations' use of phonograph records and other forms of mechanical reproduction, which would lead to a duplication of service and would not give the public anything they could not receive without radio stations. The commission encouraged stations to use local talent for original live programming. The FRC's second annual report made it clear that "the commission can not close its eyes to the fact that the real purpose of the use of phonograph records in most communities is to provide a cheaper method of advertising for advertisers who are thereby saved the expense of providing an original program." The commission issued a number of general orders that regulated the use of phonograph records. Although it never banned mechanical reproductions, specific orders stipulated that stations needed to avoid deception by making it clear to the public when they were using records.[15]

Another qualitative issue with important economic and social implications was the broadcast industry's reliance on advertising for revenue. As in the case of phonograph records, the commission did not explicitly condone the use of advertising. Advertising on the radio had in fact never been popular. As we have seen, by the mid 1920s the use of advertising had become an important means for stations to generate revenue. The commission's major concern was that "such benefit as is derived by advertisers must be incidental and entirely secondary to the interest of the public."[16]

Although the use of qualitative criteria was important in a few individual decisions, a technocratic view emerged as the dominant force guiding commission policy. Influential commissioners and powerful members of different divisions within the FRC encouraged technical standards and technocratic policies that had important implications for the overall development of radio broadcasting in the United States.

Following the advice of the institute's broadcast committee in 1928, the FRC demanded that stations follow strict engineering standards. The technical standards set by the institute committee became a central part of the regulatory policy of the commission. Stations that failed to modernize their equipment in line with "good engineering practice" had their licenses revoked. One of the most important engineering standards dealt with the allowable deviation of the transmitter signal from its assigned frequency. Frequency drift was a principal cause of heterodyne interference. In 1930, the commission believed that sufficient progress in radio equipment had occurred to enable the commission to issue more stringent regulations reducing the allowable deviation from assigned frequency from 500 cycles per second to 50 cycles per second. The new standard would increase the area generally free from heterodyne interference of a 1,000-watt station from 315 square miles to 1,500 square miles. When radio engineers from the FRC's engineering division testified as witnesses at hearings evaluating license applications, commissioners and commission staff members asked them to evaluate the technical qualifications of the station equipment.[17]

The commission's requirement that stations follow "standards of good engineering practice" by using modern broadcast equipment was an important factor in the overall decline of nonprofit, particularly educational, stations after the establishment of the commission in 1927. Especially relative to commercial stations, educational stations, mainly owned and operated by universities and colleges, lacked the financial resources to buy the expensive new equipment. A total of twenty-three educational stations went under during 1928, the first year that stations felt the full effect of commission policy; by contrast, only eight stations had failed during 1926 and 1927. The number of new licenses awarded to educational stations also fell dramatically after 1927. Whereas during the period from 1922 to 1926 the Commerce Department issued 185 licenses to educational stations, over the next seven years the FRC issued only twelve licenses. This trend led a number of critics to complain that the radio commission was biased against noncommercial educational broadcasters.[18]

During the seven-year period when the commission was in charge of regulating the industry, commercialized, corporate-controlled network radio grew to dominate radio broadcasting. Network radio did not get off the ground until the formation of the National Broadcasting Company (NBC) in 1926. The Columbia Broadcasting System (CBS) network was established in 1927 and grew quickly to become, by 1933, the largest network in the world (with ninety-one stations). By 1937, nearly 93 percent of the total transmitting power of all the broadcasting stations in the United States (a total of more than two million watts) was controlled by what one authority has called "an oligopoly of networks."[19]

Commercialization reinforced this corporate control. Despite the commission's early public stand against the use of advertising, a slow erosion of standards occurred during the late 1920s. The major networks had at first held to a company policy of prohibiting the quoting of prices on the air, but by the early 1930s companies abandoned this policy as network executives worked with advertisers to promote the use of direct advertising. Advertising agencies not only gained control of advertising policy, but also of the content of programs. By 1931, "virtually all sponsored network programs were developed and produced by advertising agencies." Critics complained that because advertisers appealed "to the greatest possible number" and the "lowest common denominator," radio broadcasts presented, "at times, a disgusting similarity in the program material."[20]

Although the commission supported these developments, neither the FRC nor Congress had ever formally taken a direct stand in favor of commercialized networks or against educational stations. An argument might be made that commercialized network radio grew to dominate radio broadcasting because the commission was a weak agency, uncertain about its role. Indeed, in 1930 the monthly publication *Radio Broadcast* called the FRC the "jellyfish commission."[21] In the beginning, the commission did not have the full support of Congress, which failed to appropriate funds. Also because of a lack of personnel, including two commissioners who died soon after being appointed, the commission at first found itself dependent on the help of Secretary Hoover and the Department of Commerce.

But the commission did actively try to regulate the industry by setting standards and reallocating stations. Robert McChesney has explored the reasons for the collapse of a fledgling broadcast reform movement in the 1930s that opposed the actions of the commission. We also need to look more closely at the technical policies of the commission and the relationship between these policies and broader social, political, and economic developments. We have examined the role of technical standards in the demise of educational stations; we also need to look more closely at other technocratic policies of the commission that had important implications for the broadcast industry. Most important, the system of allocation developed by engineers must be examined in detail. This allocation system became an important component of the large-scale technological system of communications that emerged in the United States during the 1920s and 1930s.[22]

Technocratic Means and the Allocation Plan

Developing a nationwide system of allocation was one of the first problems that the FRC tried to solve after its establishment in the spring of 1927. Significantly, the commission considered this primarily a technical problem and sought the "best scientific opinion." After initial attempts to clear up interference in some of the more congested urban areas, the experts developed a national system during the spring and summer of 1928. The FRC announced General Order 40 in August of that year. In addition to the public-interest standard, the new system was constrained by a major consideration prescribed by Congress, the Davis Amendment, which became law on March 28, 1928. The Radio Act of 1927 had authorized the radio commission for a one-year period only; the commission did not gain permanent status until December 1929. When Congress renewed the act in 1928, it added the Davis Amendment to placate members, especially from the South and West, who wanted to restrict the growing dominance of network radio and its high-power stations, partly by distributing broadcasting facilities equally throughout the country, not only to commercial stations but also to independent and educational broadcasters. The final, watered down version of the amendment, sponsored by Congressman Ewin Davis of Tennessee, stipulated only that the commission should make a "fair and equitable" distribution of facilities among the states and the five zones delineated by the commission.[23]

Dellinger argued that General Order 40 was the "closest approach to the ideal set-up which can be made at this time." Although he considered the new allocation a "compromise" between engineers' recommendations and a plan presented by broadcasters, he also boasted that the final plan was in "essential accord with the recommendation of radio engineers." The engineers' plan was based mainly on the recommendations of the broadcast committee of the Institute of Radio Engineers, which presented its results in early April at an engineering conference chaired by Dellinger.[24]

The elimination of heterodyne interference was one of the main goals of the commission. Radio engineers concluded that since the problem occurred when two stations tried to operate on the same channel, the only solution was to give powerful stations exclusive use of many of the ninety channels available in the United States (six channels were reserved for Canadian stations). An ideal setup would thus divide stations into two classes: powerful, clear-channel stations that nearly all listeners could pick up, and less-powerful, local stations sharing channels. In order to take into account the large number of stations already broadcasting, the engineers' proposal included a third class of moderately powerful stations, called "district service stations." Basing their analysis on similar technical advice that they had given to Hoover, the engineers recommended that the ninety available channels be classified into three groups: fifty national or exclusive channels for high-power stations (at 5,000 to 50,000 watts) serving all parts of the country, including rural areas; thirty-six semiclear channels for the use of moderate-power stations (300 to 1,000 watts) serving regional areas; and four local channels for low-power stations (up to 250 watts) serving small towns. The fifty national channels would be divided equally among the five zones; each zone would thus have ten national, or clear-channel, stations. Half of the thirty-six regional channels would be available for each zone, resulting in a total of ninety regional stations operating in the country. All five zones would have five stations operating on each of the four local channels, so that the country as a whole would have a total of one hundred local stations. To further satisfy the requirements of the Davis Amendment, the engineers' proposal "allotted to each state the number of assignments of each class which corresponds to the proportion of its population to the population of the zone."[25]

Representatives of the Radio Manufacturers Association, the National Association of Broadcasters, and the Federal Radio Trades Association presented an alternative plan at a meeting on April 23. Although the engineers' plan sought to eliminate many of the approximately seven hundred stations operating in the United States, the broadcasters' plan provided for their continued operation. The latter plan also limited high-power stations to 10,000 watts and did not provide authentic clear-channel stations. Dellinger called the plan a "plea for the status quo," which "reveals a serious, and almost total, lack of understanding of the import of the . . . recommendations of the engineers." The "fatal weakness" of the proposal, according to Dellinger, was its failure to consider listeners outside the local service area for each station. Only unrestricted, high-power, clear-channel stations could provide rural service. This was a fundamental consideration for the engineers.[26]

The commission continued to discuss different plans throughout the summer of 1928. In an effort to promote the engineers' proposal, Dellinger submitted detailed critical commentaries on rival plans to the commissioners. The actual system adopted by the FRC differed only slightly from the engineers' recommendations. Instead of fifty exclusive channels, General Order 40 made provision for forty. Further, instead of eliminating several hundred stations, the final proposal forced many stations, especially small independent and educational stations, to share time with others broadcasting on the same channel. The compromise also included a provision for the future elimination of any station that failed to provide proper public service. Although during the next four years the commission made some adjustments to the new allocation, their effects were "relatively small."[27]

The FRC used the allocation system developed by radio engineers as an important technical standard guiding their decisions. It evaluated applications for license renewal in terms of the quotas for different states and zones. For example, when station WSUI of Iowa State University applied for permission to operate full-time, radio engineers determined that "the granting of this application would increase the facilities of both state and zone by .34 units." Since the state was already 80 percent over quota, and the zone 27 percent over quota, the commission denied the application.[28]

The engineers interpreted the principle that officials should regulate radio according to public interest, convenience, and necessity as demanding that they provide the best service based on quantitative standards of technical efficiency. Dellinger's official report stated that the allocation system would result in "more programs at higher signal strengths by a greater number of listeners in a larger total area than at present and will do this with less interference than now exists." According to this view, improvements in the power, efficiency, and rationality of technological innovations become ends in themselves, rather than means toward achieving clearly articulated political, social, and economic ends.[29]

Engineers played a central role in the policy work of the FRC because influential members of the commission held similar technocratic views. An emphasis on expertise, instrumental rationality, and engineering language was, of course, a common theme in the political culture and social thought of the early twentieth century. The growing cultural authority of science and engineering led to such programs as the technocracy movement and scientific management. The commissioners and the legal counsel involved in important policy decisions such as the reallocation argued that broadcast regulation was fundamentally a technical problem demanding technical expertise. Key members of the commission believed they could avoid divisive political debate by implementing decisions arrived at through the use of technical reason.[30]

Louis Caldwell, the first general counsel to the commission, has been characterized as "the most important" and "the most visible legal authority on broadcast policy" during this period.[31] Caldwell served on the radio commission from June 1928 to early 1929, the crucial period when officials developed and implemented General Order 40. He was instrumental in organizing (in 1928) and then leading (until 1933) the Standing Committee on Radio Law of the American Bar Association. Both Caldwell and his committee supported the engineers' reallocation.

Caldwell's legal defense of General Order 40 emphasized that the regulation of radio broadcasting involved "highly technical problems and complicated issues of fact which are unsuited for decision by a legislative body." "In a combat between the laws of science and the laws of governments," he argued, "the ultimate victory will always fall to the former." The main example of legislative meddling for Caldwell was the Davis Amendment, which, although offering an opportunity to develop a new allocation, also imposed rigid laws on a scientific field governed by the "inexorable principle[s] of radio engineering." According to Caldwell, "it is extremely important that the future of this rapidly progressing field of science and its application to human activity be not shackled by legislation or judicial decisions resting on unscientific premises." Important members of the legal division thus contended that "radio experts" should primarily develop regulatory policies; these were experts who would provide evaluation free from the "bickering and recrimination of expert witnesses in other fields." The assistant general counsel Paul M. Segal in 1930 wrote that because "radio is essentially a scientific, not a legal enterprise, . . . the radio engineering profession must supply the impetus, the principles, and the leadership" for the development of radio policy.[32]

The contention that officials should primarily treat broadcast regulation as a technical problem provided Caldwell with a powerful argument against opponents of the policy of the FRC. Owners of small stations, independents and educational broadcasters, were especially upset with the commission's sanctioning of high-power, clear-channel stations, an action they believed created unfair competition. When groups such as the Independent Broadcasters Association and the Radio Protective Association accused the commission of disregarding the antimonopoly section (section 17) of the 1927 Radio Act, Caldwell contended that they failed to understand the "principles of engineering," which dictated high-power stations and exclusive channels as the only possible solution to the allocation problem. Another opponent, Joy Elmer Morgan, chairman of the National Committee on Education by Radio, claimed that Caldwell's support of General Order 40 and his refusal to consider reserving thirteen or fourteen channels for the exclusive use of educational stations was motivated by his close connection to the FRC and the commercial radio industry. In response, Caldwell contended that he "had no such purpose. I merely recited facts which are virtually undisputed among reputable radio engineers and which Mr. Morgan may verify by consulting the departments of physics in the educational institutions in whose name he speaks."[33]

Influential commissioners shared the Legal Division's belief in the important legitimating role of radio engineers. The commissioner who played the most significant role in the development of General Order 40, Orestes M. Caldwell (no relation to Louis Caldwell), was the only commissioner during this period who had been trained as an engineer. Commissioner Caldwell stressed that the policy work of the commission was primarily based on "sound engineering principles" and "conditions imposed by . . . stubborn scientific facts." The statements released to the press by the commissioners made a point of emphasizing that the commission had treated the reallocation as an engineering problem to be solved by independent technical experts. The FRC pointed out that "the services of some of the outstanding radio engineers of the country" had been obtained, including consulting engineers "known to be independent of connections which might in any way prejudice their views." The source of political authority for the allocation was thus closely connected to the cultural authority of technical experts.[34]

Less-influential voices on the commission opposed the FRC's technocratic policies. When Bethuel Webster became the general counsel in 1929, he criticized the commission for delegating its authority to the Engineering Division. The division, he argued, rather than serve the commission in a restricted advisory capacity, had "undertaken to give advice beyond its own field." The division was routinely forming conclusions and reaching decisions on its own, instead of supplying the commission with relevant "statements of specific facts" so that they could make policy. Webster was particularly concerned that the Engineering Division was avoiding issues of accountability by not publishing relevant engineering data.[35]

The technocratic policies of the commission assumed that officials could reject licenses based on technical or quantitative reasons that were clearly separate from sociopolitical or qualitative considerations. In practice, Dellinger drew a sharp boundary between the "technical part" of issues and other considerations that were "purely a matter of policy." When testifying at hearings on behalf of the engineering division of the commission, radio engineers sometimes refused to answer questions that they considered political rather than purely technical. They wanted to make a clear demarcation between "an opinion on the commission's policy" and "the opinion of an engineer." When asked to evaluate an issue he considered political, one of the commission's expert witnesses argued that, "as an engineer," he did not think that he was "competent to testify in that connection."[36]

By not taking into account the possibility that technical evaluations might have social, political, and economic implications, the commission was able to achieve "closure." With controversial regulatory issues, an important way officials achieve closure is through a process of reducing and narrowing the salient issue to technical, instrumental problems. This is what happened in the case of WSUI, the Iowa State University station. By basing its decision on technical considerations, the commission interpreted the "public interest" in restricted terms. Although it acknowledged the station's significant qualitative contributions to "public interest," the commission did not consider these important enough to influence the final decision. The commission conceded that WSUI was respected as "one of the pioneers" in radio broadcasting, that the station had "been unusually progressive in the promotion of education by radio," and that its current plans were clearly aimed at strengthening these "educational purposes." But the exclusively technical criteria assumed that educational stations were no more valuable than commercial stations.[37]

The radio commission's policy forced a number of educational stations and small local stations to shift repeatedly to less desirable times and frequencies. The commission gave nearly all the educational stations part-time assignments, usually during less-desirable daytime hours, which most stations considered "useless for adult education." Before the establishment of the FRC, station WCAC, run by Connecticut State College, had operated on "unlimited" time at 500 watts. During 1927 and 1928, the commission shifted WCAC two times and ordered it to share time with other stations. Over the next two years, the commission reduced WCAC's power to 250 watts, shifted its frequency, and ordered it to divide time with more stations. Interference and inadequate power were a constant problem for WCAC. In the face of constant disruptions, the college eventually abandoned its efforts and wrote the commission that "since the differences between the fundamental motivating forces in educational broadcasting and those in commercial broadcasting forever render competition on an equal basis impossible and since no federal policies have been adopted to equalize the competition for radio facilities, there is little hope that WCAC service can ever develop into a significant state educational project." The commission ordered another educational station to change frequencies eight different times, two other stations shifted seven times, and four stations shifted six times. Officials ordered Rennselaer Polytechnic Institute's station WHAZ in upstate New York to change its broadcast frequency three different times during one month.[38]

Stations competed for the best assignments. Commercial stations actively sought, often successfully, to appropriate the time slots and frequencies used by educational stations. A commercial station could request the commission to transfer for its use the time that it shared with an educational station. For educational stations, "survival seemed to require constant legal services and a budget for ceaseless travel to and from Washington." Nebraska Wesleyan station WCAJ spent years fighting against commercial station WOW's attempts to gain its time slot; in 1933, it decided against spending more time and money in "endless litigation" and sold its facilities to WOW. Critics were especially upset because although the commission had "repeatedly denied" WCAJ's request to increase its power to 1,000 watts, arguing that this would put the state and zone over quota, after the station's sale the commission granted WOW "full time on the same frequency with 1,000 watts power, thus automatically increasing the quota by .06 units, without a hearing."[39]

Technological Systems and Engineering Professionalism

An examination of individual decisions gives us only a partial understanding of the implications of the FRC's technocratic policies. We gain a better understanding of the overall effect of these policies by placing our analysis in the context of recent work in technology studies that attempts to avoid artificially imposed dichotomies by emphasizing that the technological enterprise should be understood as "simultaneously a social, economic, and political enterprise." Thus, the allocation plan based on criteria of technical efficiency and instrumental rationality was not neutral with respect to sociopolitical developments. As Langdon Winner has argued, "what appear to be merely instrumental choices are better seen as choices about the form of the society we continually build, choices about the kinds of people we want to be."[40]

The engineers' recommendations to the FRC became an important component in the system of commercialized network radio that developed during the late 1920s and early 1930s. The allocation plan accorded well with a network structure. The allocation system favored high-power, national (clear-channel) stations, used as anchors for networks operated by large commercial broadcasters. The secondary status of the local stations in the allocation plan also supported networks, which used the small stations as individual components, helping to expand the system. The allocation plan resulted in the networks gaining control of all but two of the national (clear-channel) stations, almost three-fourths of the unlimited time transmitters, and most of the regional facilities. The only commissioner to oppose the 1928 reallocation, Ira Robinson, refused to take part in its implementation on the grounds that it favored the commercial networks. Robinson was not, in principle, opposed to commercial radio; undoubtedly many of the nonprofit, educational stations that failed during this period had not been providing service of a high enough quality to justify continued licensing. Robinson specifically criticized the commission for failing to regulate network domination of broadcast radio and for refusing to control broadcasters' excessive use of advertising.[41]

The FRC served a role analogous to the regulatory agencies that supported the growth of large-scale electric light-and-power systems during the late nineteenth and early twentieth centuries. According to Thomas Hughes, these technological systems gained power and momentum by successfully controlling and incorporating the external environment, including "legislative artifacts," which became important components of the systems. Instead of seeking to adapt means to ends, technological systems exhibit what one scholar has termed "reverse adaptation." That is, a system becomes committed to molding ends that will help it maintain its momentum. The technological system in which the radio commission became a component included the commercial networks and their local affiliates, an oligopoly of radio manufacturers, professional engineers and their organizations that provided technical advice, and the values of consumer culture. The particular style of the system that developed was also congruent with broader political and economic values. Historians have characterized the decade of the 1920s in terms of consolidation and centralization. By 1929, the two hundred largest U.S. businesses (not including banks) controlled nearly half the corporate wealth in the country. The fact that the FRC supported a communications system in accord with these values should be seen in this larger context. After the business prosperity of the 1920s, the economic depression of the 1930s reinforced established trends. Many noncommercial stations failed during the early 1930s, before the rise of New Deal policies that might have provided alternative options.[42]

In order to gain a better understanding of the role of radio engineers in the social construction of radio broadcasting, we need to place the technocratic policies of the radio commission in the context of engineers' conflicting loyalties and conflicting notions of public responsibility. Historians of engineering professionalism have emphasized that during the 1920s U.S. engineers forged an alliance with the businessmen who paid their salaries. Rather than view this as an inevitable development, however, Edwin Layton and other scholars have emphasized that the "origin of engineers carries with it built-in tensions": engineers need to deal with conflicting commitments not only to the well-being of the public but also to the interests of the corporations or bureaucracies and to the independence or autonomy of their profession.[43]

In a limited sense, we see how different resolutions of engineers' dilemmas during the late 1920s and early 1930s resulted in divisions within the Institute of Radio Engineers. A few members emphasized both the importance of professional independence and an activist definition of engineers' social responsibility and commitment to the public. At least two radio engineers criticized the commercialization and dominance of network radio and emphasized their profession's need to take responsibility for the development of radio broadcasting. Lee de Forest, whom some people refer to as the "father of radio" because of his "invention" of the triode vacuum tube, became president of the institute in 1930 and used this position of authority to campaign for the reform of radio broadcasting. In both of his presidential addresses to the institute, he spoke out against what he called "the greed of direct advertising," and he wrote letters of protest to government officials, newspapers, radio-trade journals, and general periodicals. In 1932, he lamented that "within the span of a few years we in the United States have seen broadcasting so debased by commercial advertising that many a householder regards it as he does the brazen salesman who tried to thrust his foot in at the door."[44]

De Forest also believed that engineers should play a central role in reforming radio broadcasting. He felt it was their professional "duty" to—in his own words—"take active steps (in Washington if need be) to rid ourselves of this [system]." Both de Forest and Stanford Hooper, the most important radio engineer in the U.S. Navy, called on engineers (especially members of the Institute of Radio Engineers) to take responsibility for the implications of their work. Hooper had played a key role in the development of national radio policy. Perhaps most important, at the end of World War I he had helped organize the Radio Corporation of America (RCA) as a U.S. company that would serve the national security needs of the federal government, especially the navy, and he also served as an adviser to the FRC. Hooper felt strongly that engineers should never lose sight of their loyalty to the public. He called for engineers to assume responsibility as "standard bearer[s] to speak to and for the public as to what is best for the latter's interests and those of the nation in radio, and to represent the public before the government." The fact that they had not been doing this, he considered to be "an indictment of the radio engineer."[45]

But the views of Hooper and de Forest did not represent the views of many other radio engineers, especially those involved in the regulation of radio broadcasting. Hooper himself, however, had advised the commission about regulating high-frequency radio, not "popular" radio broadcasting, and although at times he expressed sympathy with the aims of the broadcast-reform movement, he generally supported strong corporate control for the radio industry. De Forest, on the other hand, despite becoming president of the Institute of Radio Engineers in 1930, was more of a maverick outsider than a representative leader of the engineering community.[46]

Of course, many—perhaps all—of the radio engineers who advised the radio commission about the regulation of radio broadcasting also felt a sense of loyalty to the public. Their loyalty similarly stemmed from a confidence in their specialized, esoteric training, which they believed could help solve social and political, as well as technical, problems. Dellinger—who, as we have seen, was one of the most important consultants on national radio policy—expressed this technocratic sentiment when he argued that the public should not view the engineer as "a mere tender of machines, as the world rapidly becomes a civilization of machines, the masters of machines will increasingly be the ones in control of the world." Dellinger believed that so far as national radio policy was concerned, "it is the radio engineer who forges the keys to radio paradise for an ever-growing number of people."[47]

But the radio engineers responsible for the technical policies of the radio commission tended to interpret social responsibility in terms of business interests. Before the early 1920s, radio engineers and government regulators generally tried to draw a sharp distinction between the professional, "scientific" concerns of the Institute of Radio Engineers and the corporate interests of the major employers of engineers. Early members of the institute after its founding in 1912 included several important members of the American Institute of Electrical Engineers who had "transferred their activities" from the former organization to protest its control by big business. When the institute provided technical advice to the Department of Commerce concerning regulations proposed at the interallied commission in Paris, the president of the institute wrote the secretary of commerce that "we find it impossible to express in a Committee Report the views of any corporate or other radio interests as distinguished from recommendations conducive to the best progress of the radio arts and practice."[48]

During the 1920s, this clear separation (or at least perceived clear separation) between engineering independence and business loyalty tended to break down. The engineers who played major roles in advising the government about regulating radio broadcasting developed close ties to large corporations, which became increasingly important as the major employers of radio engineers. Of the seven most important radio engineers (excluding Dellinger) who influenced the policies of the radio commission, three were independent consulting engineers: C. M. Jansky, Robert Marriott, and John V. L. Hogan. But the degree to which they were all acting independently was by no means clear. Hogan both advised the commission about the need for more clear-channel stations and worked (along with Louis Caldwell) for a group of broadcasting stations lobbying the commission for this same policy. The other four engineers worked for major communications corporations and radio manufacturers: Laurens Whittemore and Lloyd Espenschied were employed by AT&T, C. W. Horn was an employee with Westinghouse, and Alfred Goldsmith worked for both RCA and NBC.[49]

Goldsmith's influence was especially important. He had been actively volunteering to help the Department of Commerce with problems in radio regulation, especially "the allocation of wavelengths to prevent interference," since at least 1917. Some of the policies he helped establish during the mid 1920s, such as the emphasis on high-power and clear-channel stations, served as a basis for the policies of the FRC. Although he was a founding member and a central leader of the Institute of Radio Engineers, his interest in public policy for radio broadcasting was also motivated by the corporate concerns of RCA and NBC. Goldsmith became chief broadcast engineer of RCA in 1923 and, in 1927, chairman of the board of consulting engineers of NBC. As early as 1922, executives at RCA requested that Goldsmith volunteer his services to the Department of Commerce in "forming any new rules or regulations which shall govern the assignment of wavelengths." Further, Goldsmith served as a chairman of an RCA committee that concluded that advertising was the only viable means to finance radio broadcasting. Goldsmith's influence on the FRC also reflected corporate commitments. At Goldsmith's request in December 1928, the broadcast committee of the Institute of Radio Engineers agreed to hire "somebody from the Columbia chain." Despite evidence of an accommodation between radio engineers and business interests during the 1920s, Goldsmith claimed publicly, in 1925, that "the Institute of Radio Engineers is an entirely non-partisan and scientific society, free from any commercial affiliations or connections whatsoever."[50]

Although Dellinger worked as a government employee, he was considered sympathetic to commercial interests. For these reasons, he was chosen, in 1922, to replace another employee also working in radio regulation at the Department of Commerce. Dellinger's close ties to commercial radio partly reflected Hoover's influence. Under Hoover's direction, the Department of Commerce sought to assist business growth and consolidation. The engineers in the department were responsible for rationalizing radio broadcasting by coordinating business interests. The close connection between business executives and professional engineers is evident in the membership of the technical group organized in 1926 to assist the Department of Commerce with the policies of radio regulation. The radio broadcasting committee of the American Engineering Council included among its members David Sarnoff, the vice president of RCA.[51]

Evidence of institutional accommodation gives us only a partial conception of the relationship between the technical policies of the radio commission and broader political, social, and economic developments. We also need to understand how the particular values embedded in the technical policies of the commission were congruent with the values of 1920s corporate America. Both engineers and business leaders shared a commitment to centralize, gain control, and impose order. Both groups were responsible for constructing the large-scale, integrated technological systems of the twentieth century.

The leaders of the radio engineering community generally shared the laissez-faire attitudes of business leaders. Since the first debates about the regulation of radio, engineers had spoken out against excessive government intervention. This was especially true before the rise of popular radio broadcasting in the 1920s. Goldsmith argued that since interference was a technical problem, it could "not be legislated out of existence by the extinction of healthy development, but avoided by sound engineering expedience." Engineers believed political solutions would only have a "retarding effect upon the technical and commercial development of radio communication." Thus, technological problems such as radio interference could be solved only through further technological development. Although the leaders of the radio engineering community generally did not want government intervention, they did approve of government regulation that supported business involvement, especially in the form of technical support from the new industrial research laboratories. Michael Pupin wrote in 1917 that "things are being done today by well organized industrial research laboratories which will undoubtedly lead to wonderful results so far as preventing interferences produced by the acts of man are concerned." The leaders of the radio engineering community in the 1920s continued to view corporations as primary agents of technological growth. The radio engineers who advised the radio commission preferred an economic system based on commercial advertising partly because they believed this would best ensure limited government involvement and unrestricted technological development in industrial research laboratories.[52]

Engineers shared business executives' commitments to pragmatic values such as efficiency and planning as well as to the centralized control necessary for the construction of technological systems. Both groups were committed to large-scale development, both technological and organizational. C. W. Horn, who advised the radio commission while serving as superintendent of radio operations at Westinghouse, argued that "the public wants the big features that occur, wherever they occur." Alfred Goldsmith not only oversaw the construction of high-power network radio at NBC and helped shape commission policies that encouraged these developments, but also actively opposed legislation that included provisions against monopolies. He argued that neither the secretary of commerce nor the FRC should be given the power to decide the fate of large businesses.[53]

An emphasis on high-power stations and exclusive or clear channels were the two most important provisions of General Order 40. The engineers who developed General Order 40 argued that the allocation was demanded by rigid engineering principles. High-power stations were part of the "inevitable process of evolution in radio communication" and a "symbol of the progress of the science." According to one proponent, "it would be the greatest kind of a mistake to meddle with a pioneer art or to attempt to stop scientific and technical development." But the allocation was also premised on a central goal that RCA had been emphasizing since its first experiments with high-power stations in the early and mid 1920s—the need to provide high-quality service to the country's rural areas. In 1925, while he was chief engineer at RCA responsible for the construction of high-power stations, Goldsmith asked the secretary of commerce to consider "how many hundreds of thousands of unfortunate people in the United States today, situated far from the high-grade metropolitan stations, are . . . being deprived of the to them irreplaceable benefits of radio." A critical view, which underscores the need to evaluate technological developments within their appropriate historical context, including the particular values of its practitioners, was expressed by Commissioner Ira Robinson, who argued that the radio engineers who "have testified that high power is the thing . . . are naturally by their employment big business minded." They were not necessarily "biased" or dishonest in a simple-minded or crude sense of those terms. Because they generally did not focus on broad issues, if anything they were guilty of lacking a sense of self-criticism and the ability to be deeply reflective.[54]

By exploring the interaction of technology and society and the ways in which particular attitudes and world views are embedded in engineering decisions, we gain a better understanding of why particular technological developments occur as they do. We also see how successful engineers in the twentieth century have been involved in the creation of technological systems; they not only invented individual artifacts but also essential social, economic, and political components. The central role of radio engineers in the policy work of the FRC underscores the importance of technical experts for political decision making. But radio engineers were not simply usurping political authority from politicians and administrators; nor were they simply being used by others to advance political agendas. We need to recognize the symbiotic relationship among the members of the different groups involved in radio regulation. Politicians and administrators delegated authority to engineers, but the engineers shared with them certain values. These values, which were embedded in their technical decisions, embodied essential characteristics of the political, economic, and business climate of the 1920s.

The social shaping of broadcast technology should thus be seen as part of a complex reciprocal process that reflects the interactive nature of technological change and social development. The technical policies made by technical experts played a major role in shaping the technological system of radio broadcasting in the United States, but these decisions were themselves shaped by social, political, and economic realities that became subsumed within the system. It is important to emphasize that the developments detailed in this chapter were made possible by the growing cultural and social authority of science and engineering during the 1920s and 1930s. There was a widespread public perception that scientific and engineering decision making was "objective" and "neutral." Participants in the process therefore delegated authority to experts as a way of rationalizing public policy. In 1934, the FRC gave place to the FCC. This process of delegation to experts was to continue, but a different political climate and new technological developments led regulators to search for new ways to institutionalize the role of expertise.

3)))

Competition for Standards

Television Broadcasting, Commercialization, and Technical Expertise, 1928–1941

The real possibilities of television as a means of education and entertainment are today totally unknown. [It will] probably have more effect on the life of the American people than any system known today.

W. R. G. Baker, chairman of the
National Television System Committee,
July 31, 1940

Television and FM radio, the two major broadcast technologies first introduced during the 1930s and 1940s, presented government officials with new policy challenges. This chapter discusses the events leading to the commercial authorization by the Federal Communications Commission of electronic television in 1941 (the FM development is discussed in chapter 4). World War II delayed television's emergence as the dominant mass medium in the United States, but the federal government and the television industry made many of the fundamental policy decisions before Pearl Harbor. By this date, ten commercial television stations were broadcasting and approximately ten thousand television sets were in the hands of the public.[1]

Beginning in the 1930s, the Federal Communications Commission (FCC) actively sought to oversee television's development. The dramatic growth and success of radio broadcasting following the first broadcasts of KDKA had caught most of the country by surprise. Television broadcasting, by contrast, had been anticipated for many years. The two different regulatory histories also reflected different economic and political climates: whereas Hooverian Associationalism influenced the development of radio during the 1920s, the effort to regulate more closely the commercial growth of television reflected the activist policies of New

Deal decision makers. In planning for the development of television, government officials sought to avoid some of the ad hoc decision making made for radio during the 1920s. The FCC placed a strong emphasis on using formal rule-making proceedings to develop a master plan for television. Rule making is an essential function of regulatory institutions. The process allows agencies to carry out the implications of statutes passed by Congress. The rule-making process, which involves soliciting responses from industry and government officials as well as citizens groups about proposals, gives legitimacy to decisions and provides for rational and uniform planning, but it also can lock officials into a rigid system not easily adapted to changing conditions. As it developed in the effort to establish a nationwide system of television stations (see esp. chapter 5), it can also be a lengthy and complex process, involving numerous meetings and the evaluation of many layers of comments and replies to comments made by outside groups about different reports and orders proposed by the commission.[2]

As with radio, issues involving engineering standards became the major focus of concern for government officials attempting to oversee the development of television, but a fundamental difference helped define new problems. During the 1920s, the crucial issue involving technical considerations for both the Department of Commerce and the FRC had been the determination of an ideal assignment system for radio broadcasting that would reduce interference between stations. In the case of television, most of the controversy over spectrum allocation would wait until after World War II, when there was the opening up of the ultrahigh frequencies (UHF); in the decade before the war, the contentious problem for the FCC was the authorization of a complete technical system for electronic television. Regulators emphasized that this problem distinguished television from other industries, and notably from radio broadcasting. Unlike radio, television demanded a "lock-and-key" system, in which the operation of receivers was synchronized, with a high degree of accuracy, with broadcasts from compatible transmitters. The receiving set would unlock the transmitter to receive the broadcast.

The FCC refused to authorize commercial television broadcasting until industry and government engineers agreed on technical standards for all transmitters and receivers. This chapter thus analyzes the involvement of different individuals and institutions in the establishment of system standards and the authorization of commercial television. Here again, recognition of the tendency toward technocratic decision making helps us to gain a better understanding of this development. Although a "technical" problem based on physical principles and equipment characteristics, the determination of system standards also involved important economic, social, and political considerations. An analysis of the advisory role of engineers and the strategies used by participants in the standards debates is particularly important for understanding how policy makers resolved fundamental problems connected to the theme of technocracy. We will especially be interested in exploring how the FCC tried to manage crucial issues by experimenting with new arrangements to institutionalize the advisory role of technical experts. Before addressing these questions, however, we first need a brief account of the early development of television.

The Early Years of Television and the Role of RCA

By the mid nineteenth century, a number of inventors had proposed using radio waves to transmit images. The discovery, in 1873, that the electrical resistance of selenium decreased when experimenters exposed the element to light inspired specific proposals involving the use of a series of separate photosensitive cells. When an experimenter exposed light from an illuminated object on a selenium mosaic, each cell would draw a current proportional to the brightness of the part of the image illuminating it. The major problem with this proposal was that its reliance on different circuits for each element of the picture would demand more space in the electromagnetic spectrum than was available. Alternative techniques sought to scan, and then project using one channel, all the parts of an image at one time at a speed fast enough to take advantage of the inability of the human eye to distinguish between rapidly projected visual images. Experimenters would scan the sequence of pictures and transmit them at a sufficient rate so that this "persistence of vision" of the human eye would give the illusion of fluid motion. The German researcher Paul Nipkow developed one of the most important early systems of "sequential scanning." First proposed in 1884 and developed by other inventors during the next three decades, his mechanical system projected light from an image first through small holes arranged in a spiral on the outside face of a thin disk and then onto selenium cells. With one rotation of the disk, each hole scanned a different part of the image; this technique resulted in multiple scanning lines, or one complete scan of the entire image. The system transmitted the changing electrical signals from the selenium cells through one channel to a receiver, where it powered a light source. Using an identical revolving disk, synchronized with the one in the transmitter, an experimental apparatus would project the modulated light source onto a television screen.[3]

Inventors succeeded in presenting public demonstrations of mechanical television during 1925 on both sides of the Atlantic Ocean. Historians generally credit the American Charles Francis Jenkins and the Briton John Logie Baird with the invention of a workable system based on the original Nipkow proposal. During the late 1920s, American Telephone and Telegraph Company (AT&T), General Electric (GE), and other large communication-equipment manufacturers in the United States made further improvements and conducted more spectacular demonstrations, mainly to the press, of entertainment as well as news and sporting events. A number of radio stations also began experimental television broadcasts to the public during this period. Spurred on by these new developments, for five years, beginning in 1928, a new industry based on mechanical television flourished in the United States. The FRC issued twenty-one television licenses during 1928. Statistics on the number of television sets in use during this period are limited, but one observer estimated that, by 1931, in Chicago consumers owned eight thousand television sets. The failure of mechanical television to live up to expectations of high-quality images comparable to film presentations contributed to the collapse of the industry in 1933. Radio survived as a young and still imperfect technology by relying on its communication role; television's early development, by contrast, depended on cultivating an entertainment role, which demanded higher standards of quality. The deepening of the Great Depression during the early 1930s also stifled interest in television research.[4]

Despite the failure of mechanical television to sustain public and corporate interest, the early experiments with manufacturing and broadcasting helped lay a foundation for a new industry based on an alternative method of televising images: all-electronic television. Researchers first became interested in the principle behind electronic television during the 1850s, when they observed the behavior of electrical discharges between oppositely charged plates in a vacuum, or "cathode ray," tube. After the turn of the century, inventors proposed specific designs for television receivers based on an earlier observation that the "cathode ray" discharges caused fluorescence on the tube's walls. By focusing and directing the electrons using electromagnets or electrostatic devices, inventors sought to use the cathode ray as a scanning device for projecting a transmitted image onto a fluorescent screen.

A. A. Campbell Swinton made the most important early proposal for a complete electronic television system, in England, just prior to World War I. He suggested a system of sequential scanning using synchronized electron-scanning beams in the transmitter and receiver. The beam in the transmitter, or camera tube, would scan a mosaic of photosensitive cells and impart a negative charge to each cell it scanned. Cells would then react differently depending on the amount of exposure to light from the image; the system would record and transmit this response to the receiver. The scanning would move across the horizontal lines of cells, starting with the top row and moving down, scanning each row in turn. One horizontal picture line would thus represent a row of as many as several hundred different cells. The system would scan the entire mosaic once every one-tenth of a second. Like mechanical television, the electronic system would transmit information over one channel and take advantage of the principle of persistence of vision to reproduce fluid motion.[5]

Despite what supporters believed were its obvious advantages, including speed of operation and lack of moving parts, all-electronic television had to wait until the 1930s to compete successfully with the mechanical version. Experimenters needed to perfect the cathode-ray tube, the photosensitive mosaic, and other electronic elements. Manufacturers and broadcasters learned from the experience with mechanical television the importance of first developing more fully the new television system, including both engineering and programming, before attempting to sell it to the public and to advertisers.

By the end of the 1930s, thanks especially to the important engineering research of Vladimir Zworykin, the Radio Corporation of America (RCA) had become the dominant company in television research and experimentation. Zworykin emigrated to the United States in 1919 from Russia, where he had worked with Boris Rosing at the Technological Institute of St. Petersburg. By 1912, Rosing had completed significant research using a cathode-ray tube as the basis for a crude television system, but World War I and the Russian Revolution put an end to this line of inquiry. Zworykin worked as a special military engineer during the war and sided with an anti-Bolshevik faction after the revolution. He fled to the United States to avoid civil war and to continue to pursue his engineering interests. His first employer in the United States, the Westinghouse Electric Company, gave only limited support to his research on electronic television, beginning in 1923. Rosing and other European experimenters had previously demonstrated the use of cathode-ray tubes for television receivers; Zworykin sought to build an all-electronic system that also used a cathode-ray tube for the "pickup tube" in the camera. In 1925, Zworykin demonstrated his first system, which was similar in design to Campbell Swinton's earlier proposal. One important innovation was the use of charge storage in the photosensitive mosaic. The individual cells in the mosaic would hold their charges after being exposed to light from the image until the camera tube was ready to convert them into an electrical signal to be transmitted. Although observers complained about the quality of the images from this first demonstration, the event marked the first time experimenters successfully made a television broadcast with a transmitter and a receiver both using cathode-ray tubes.[6]

Over the next three years, Westinghouse failed to pursue further research based on Zworykin's experiments, choosing instead to investigate mechanical television, which seemed to hold more promise for short-term commercial success. Company executives ordered Zworykin to work on other projects not connected with his television experiments. Had David Sarnoff, the executive vice president of RCA, not taken an interest in Zworykin and decided to support his work, electronic television would probably have been delayed. Sarnoff first became interested in strongly promoting television in 1927 after viewing a successful public demonstration of a mechanical system developed by AT&T. Since he viewed television technology as properly an extension of radio broadcasting, he was anxious to prevent the telephone company from gaining any commercial advantage. Sarnoff recognized the commercial possibilities of electronic television and, beginning in 1928, made sure Westinghouse allowed Zworykin to continue his research. RCA's role was important because the agreement among the members of the Radio Group (chapter 1) stipulated that Westinghouse would conduct radio and television research to assist RCA in selling the other companies' products. The U.S. Patent Office awarded Zworykin an important patent in 1928 for his all-electronic system. During his last year at Westinghouse, he perfected and then successfully demonstrated to the 1929 meeting of the Institute of Radio Engineers a superior cathode-ray receiver; he called it a kinescope (from the Greek *kinein,* to move, and *skopein,* to watch).[7]

Beginning in the late 1920s, Sarnoff sought to reorganize RCA and renegotiate the agreement with the other members of the Radio Group. He primarily wanted RCA to have the freedom to expand into manufacturing and research, but in order to do this the company would have to gain independence from Westinghouse and General Electric, the senior partners responsible for these activities. As part of a new agreement, Westinghouse and GE transferred much of their production and research staff to a new company headed by Sarnoff, RCA Victor. The new business combined resources from the other two corporations with production facilities of the recently acquired Victor Talking Machine Company. As part of the transfer, in 1930, Zworykin became director of television research at the RCA Victor Electronic Research Laboratory in Camden, New Jersey. During that same year, Sarnoff became president of RCA. Within two years, RCA achieved complete separation from Westinghouse and General Electric, thanks to the assistance of an antitrust suit filed by the U.S. Justice Department in 1930. The members of the old Radio Group would now have to compete fully in the manufacture and sale of electrical equipment, including radio and television apparatus. RCA benefited more than the other companies from this new arrangement, and under Sarnoff's leadership, and despite the Great Depression, invested heavily in television research.[8]

After moving to RCA, Zworykin immediately set to work perfecting his design of a cathode-ray camera; in the fall of 1931, he developed a model of what he called an iconoscope (from the Greek *icon,* image, and *skopein,* to watch). His kinescope and iconoscope designs provided the foundation for RCA's first successful television system. The company conducted field tests using these two inventions in Camden in 1933. Sarnoff had at first considered initiating broadcast service during that year, but he recognized that the system needed further development and sent Zworykin and his team of engineers back to the laboratory for another three years.

RCA demonstrated the improved system to the press in April 1936. The new system scanned 343 lines per frame and transmitted 30 frames per second. A technique called "interlaced scanning" produced a high-quality picture without increasing the bandwidth by alternately scanning half the lines in the picture frame and sequentially transmitting the fields of odd numbers and even numbers to reproduce one frame. RCA conducted further tests later that year in New York City using a transmitter owned by the National Broadcasting Company (NBC) on the Empire State Building and a newly constructed studio at Radio City Music Hall. The company distributed at least seventy experimental television sets with small screens to RCA employees around the city. One observer reported enthusiastically in *Broadcasting* magazine that the RCA television demonstrations "surpassed anything I had ever seen before." He predicted that "two more years—possibly less" and the iconoscope would join the microphone "as the standard equipment of American broadcasting." Experimental broadcasts enabled the engineers to continue to improve the camera tube and receiver. Industry officials also worked to gain the interest of potential advertisers and solve the problem of creating station networks. In his 1935 announcement of the start of field tests, Sarnoff had emphasized that he did not want to begin commercial broadcasting prematurely and "freeze the art" at a low standard of quality. According to Sarnoff, this "would prevent the free play of technical development and retard the day when television could become a member in full standing of the radio family." Sarnoff emphasized the "lock and key" feature that distinguished television from radio:

> Any old sound receiver, even the amateur crystal sets of years ago, can still pick up programs from any standard transmitting station. Thus sound sets do not become obsolete. Unless they are fully standardized television sets will. The reason is this: In television every receiving set must be perfectly coordinated as to number of lines, method of scanning, size of picture and synchronization of signals with the transmitter at the broadcasting station, or else it will not work. This means, as you can see, in case a few million receivers are sold to the public, that the television art is frozen to their capabilities because any material change in the system, due to new discoveries, would instantly render all the receivers inoperative. The manufacturers of television equipment, therefore, must be absolutely sure they are right before they can go ahead.[9]

But after two years of field tests, which included experiments with different kinds of programming from the Radio City studio and the development of a higher-quality system that scanned 441 lines per frame, Sarnoff decided to push for commercial service. He announced in October 1938 that RCA would begin television service to the public at the opening of the 1939 New York World's Fair, aptly titled "The World of

Tomorrow." Eventually, he needed to convince the communications commission to change regulations that allowed experimental broadcasting only so long as a company did not benefit financially, including profiting from sponsored programs. Instead of expressing concern about potentially freezing the art at a low level, Sarnoff now argued that "if action were deferred to await near perfection, the medium would be postponed almost indefinitely."[10]

Early Competitors in the Television Industry

RCA's decision to pursue commercialization partly resulted from developments that threatened the company's competitive position. Although RCA had played a leading role in the development of electronic television, by 1939 a number of other individuals and corporations had been undertaking important research and were preparing to market television receivers and establish regular television broadcast service. Two companies in particular, Farnsworth Television, Inc., and the Philco Radio and Television Company, became increasingly successful during the late 1930s, competing with RCA in the development of electronic television.

Scholars usually credit Philo Farnsworth, the inventive force behind Farnsworth Television, as a co-inventor, along with Zworykin, of electronic television. A number of individuals had made significant contributions, but none more than either of these two innovators. Unlike Zworykin, Farnsworth was largely self-taught. He became fascinated with the idea of television after reading a magazine article while growing up on a ranch in Idaho. As a high-school student during the early 1920s, he began to think about designing his own system. By 1926, Farnsworth had convinced California investors to support his work on all-electronic television. He intended from the beginning not simply to develop a marketable system but also to acquire patents that he could use to fund further research. Farnsworth first gave a public demonstration of his system in 1928 at his new laboratory in San Francisco. Unlike Zworykin, Farnsworth did not have the support of a large corporation with deep pockets (despite the Great Depression, by 1939 RCA had invested nearly $10 million in electronic television).[11] With costs mounting and further research still necessary, Farnsworth agreed to a partnership with one of RCA's main rivals, the Philadelphia-based Philco Radio and Television Company. By gaining access to a television system different from the one RCA was developing, Philco hoped to avoid the patent arrangement it was locked into with radio—having to pay royalties to RCA for every receiver it sold.

After two years, however, Philco and Farnsworth parted company. Philco expected more progress from Farnsworth on a marketable system; Farnsworth was willing to delay production in order to perfect a system and establish a patent structure. With new support from private investors, Farnsworth set up his own research facilities in Philadelphia and, by 1936, partly in response to RCA's public demonstrations, began to push for commercial authorization of his own system. In 1937, using a new transmitter and a production studio outside Philadelphia, Farnsworth demonstrated images of 441 scanning lines transmitted at 30 frames per second. To compete with RCA and other large firms, Farnsworth's investors helped establish a new company, the Farnsworth Television and Radio Corporation, in February 1939. Farnsworth served as vice president and director of research. The company used the profits from radio and phonograph production to support further development of Farnsworth's television system. Farnsworth's strategy of developing patents to maintain competitive advantage paid off during that same year when RCA grudgingly agreed to pay a license fee for the patent on his camera pickup tube, known as the image dissector. Although Farnsworth's invention did not have the sensitivity of the iconoscope, RCA engineers had used design elements of the image dissector to make refinements on Zworykin's pickup tube: RCA's plans for marketing television service could not go ahead without Farnsworth's patent. RCA maintained a standard policy to purchase patent rights rather than pay royalties or licensing fees, but when Farnsworth held out, RCA acquiesced: they agreed—for the first time in the history of the company—to pay a fee.[12]

Philco, however, having already established itself as a leading manufacturer of radio receivers by the early 1930s, presented a potentially more formidable challenge to RCA's effort to develop and market electronic television. Until radio receivers switched to the use of AC current during the late 1920s, Philco (known until 1932 as the Philadelphia Storage Battery Company) had specialized in the manufacture of radio batteries. The firm then decided to manufacture and sell its own receivers, which meant paying licensing fees to use RCA's patents. By specializing in radios mounted in attractive wooden cabinets, Philco outsold RCA during a large part of the 1930s. With assistance from Farnsworth, the company began to operate its own experimental television station in Philadelphia in 1932. Television research continued after Farnsworth resigned, thanks to the assistance of the head of RCA's advanced-development program at Camden, who left to become Philco's chief television engineer. This gave Philco full knowledge of RCA's television research and helped fuel a growing hostility between executives in the two companies.[13]

Following RCA's public demonstration in 1936, Philco held its own demonstration. Witnesses of both events argued that since the results seemed "practically identical, . . . both companies must be working along practically identical lines." RCA, however, was in a stronger position with respect to control of patents. Philco had the freedom to conduct experimental demonstrations without worrying about patent interference, but in order to market its system, the company first had to acquire patent rights. Recognizing this state of affairs, the president of Philco argued unsuccessfully, in 1936, for the formation of a patent pool for the television industry "to avoid" both "chaos" and "monopoly." Philco became RCA's main competitor in television research and development during the following year when it introduced a system transmitting 441 lines per frame. The competition also became increasingly nasty that year when Philco filed suit against RCA, accusing the firm of conducting unethical trade practices, including obtaining "confidential information" about Philco's television work from some of its female employees by placing them in "compromising situations." A second suit sought to prevent RCA from blocking Philco's access to patent licenses.[14]

Although Farnsworth and Philco became RCA's main rivals for television development during the 1930s, by the time of RCA's 1938 announcement that it planned to establish television service, a number of other companies had also made preparations to compete. In fact, two smaller companies—Communicating Systems, Inc. (later known as the American Television Corporation), and Allen B. DuMont Laboratories—had already announced in the spring that they would begin to sell television (TV) receivers so the public could receive transmissions from RCA and other experimental broadcasters.[15]

Following RCA's announcement, the weekly newsmagazine *Broadcasting* reported that "all but a few of RCA's fifty receiving set licensees and its thirteen tube licensees indicated they would enter the field." The licenses RCA issued to the companies allowed them to manufacture television apparatus. Nine manufacturers in particular—DuMont, GE, Philco,

Westinghouse, Pilot, Garod, Meissner, Stewart-Warner, and Stromberg-Carlson—promised to have receivers ready to sell to the public following the 1939 World's Fair. Despite its important early involvement, Farnsworth Television did not have the financial resources to make a commitment to manufacturing at this time. Also notably absent from this list of companies was Zenith Radio. The FCC had recently granted Zenith a license for an experimental television broadcast station. The company believed it had the economic and material resources to start manufacturing receivers but refused on the grounds that television needed further improvement in order to avoid freezing the art at a low level. Zenith thus strongly opposed RCA's 1938 announcement and Sarnoff's evaluation that "if action were deferred to await near perfection, the medium would be postponed almost indefinitely." Zenith's president, Eugene F. McDonald (the maverick who had challenged Hoover's authority during the mid 1920s), warned that "such premature introduction of television commercially will result in loading the public with undue experimental replacement cost which, in turn, will result in retarding, instead of furthering development and in unprofitable operations for the companies engaging in such a program."[16]

But competition did not come only from manufacturers; broadcasters also prepared to compete with RCA and its broadcast subsidiary NBC. Many of the potential competitors represented experimental stations from the early years of mechanical television that had made the transition to the new electronic era. One of the most important was Columbia Broadcasting System's station in New York City. Although the Columbia Broadcasting System (CBS) had actively pursued broadcasting during the first television boom, the company delayed conversion to the new system. In December 1935, CBS hired Peter Goldmark, a Hungarian-trained television engineer, and placed him in charge of upgrading Columbia's television broadcast apparatus. As NBC's chief rival, the CBS network recognized the importance of keeping up with the new development. At first, CBS planned to buy transmitting equipment from RCA, but by 1938 Goldmark had convinced CBS executives not only to compete with RCA/NBC in broadcasting but also to try to develop an alternative technical system.[17]

Although most of the experiments with television broadcasting during the late 1930s occurred in New York and Philadelphia, there was also activity in other major cities, including Kansas City (First National Television, Inc.—active until 1941) and Boston (General Television Corp.—

active until 1945). More significant, however, were the broadcasts in Los Angeles of the Don Lee Broadcasting System. The Don Lee System made the switch to electronic television during the early 1930s, with the assistance of a television engineer trained in Farnsworth's laboratories. After extensive experimentation with different kinds of programming, by 1939 the company was transmitting for seven hours every day of the week except Sunday from a new studio and transmitter near Hollywood.[18]

Hollywood movie studios also began to take an interest in television broadcasting, especially beginning in 1938 when Paramount Pictures came to a financial agreement with DuMont Laboratories to support its development of cathode-ray tubes for use in television receivers and cameras. Television broadcasters experimented with film broadcasting prior to other formats, and a number of individuals speculated that theaters would become major users of the new technology by projecting television broadcasts on large screens. Paramount was uncertain about the future relationship between television and the film industry, but it wanted to have the freedom to pursue new possibilities.[19] With financial support from Paramount, DuMont worked on developing a unique television system, different from the one being promoted by RCA.

Federal Regulation Prior to 1939

By 1939, television seemed ready to take off. The public was interested; the technical, economic, and programming elements of the system appeared at least well under way; broadcasters and manufacturers, especially NBC and RCA, were pushing for commercialization. But ultimate legal authority for development of television as a public service rested with the Federal Communications Commission. The federal government's role in regulating television had been established during the tenure of the FRC, from 1927 to 1933. In response to the growth of mechanical television, the commission had decided on three kinds of issues: allocation of parts of the frequency spectrum and assignment of bandwidths for television service, licensing and assignment of specific stations to these frequencies, and whether to allow commercialization using authorized technical standards or limit stations to experimental broadcasting.[20]

The first television broadcasters transmitted in frequencies that standard AM radio (550 to 1,500 kHz) already used. By the early 1930s, the FRC had opened up channels in the largely unused higher bands above

40 MHz. Engineers discovered that the crowded lower bands were unsatisfactory because of the need for large bandwidths to transmit high-quality images and because television transmissions interfered with other broadcast services. Interference was less important in the higher frequencies because transmission occurred through ground waves moving in straight lines rather than from waves reflected off the ionosphere.[21]

The allocation decisions before World War II had important implications for the television industry, especially with respect to other services such as FM (frequency modulation) radio. In general, however, the problem of authorizing commercial operations and technical standards led to more significant disagreements and controversies during this period, partly because these issues involved an evaluation of the quality of television broadcasts and related economic and social concerns. As early as 1928, the radio commission established a policy of refusing to authorize any commercial use of television until the industry had agreed on technical standards that gave high-quality television images. The commission was especially concerned that it not endorse a system that would soon become obsolete. Manufacturers might benefit from this arrangement, but until they established a satisfactory standard, consumers would have to purchase new models. When evaluating applications for TV station licenses, the commission mainly took into account the quality of technical rather than programming experimentation and research. Officials clearly rejected commercial considerations, even the "free rebroadcast of commercials from radio." The 1934 Communications Act established legal authority for the FCC to continue to regulate technical standards for broadcasting. Section 303 authorized the commission to "from time to time as public interest requires, . . . regulate the kind of apparatus to be used with respect to its external effects and the purity and sharpness of the emissions from each station and from the apparatus therein."[22]

A special committee of the Radio Manufacturers Association first examined the problem of developing industry-wide standards for television in 1928. Some of the recommendations of the first Television Standardization Committee, including left to right and top to bottom scanning, did set important precedents. But the deliberations did not take into account the use of cathode-ray tubes and failed to consider the method of synchronizing the transmitter with the receiver, an essential technique to ensure the lock-and-key relationship. Further, instead of recommending one transmission standard, the committee recommended two, 48 lines per frame and 60 lines per frame (both at 15 frames per second). Although the FRC expected industry technical experts to develop standards, it made the final decision of whether they were of high enough quality to be in the public interest. The commission rejected the 1928 recommendations as a basis for a commercial system for failing to meet this test. In 1932, the television committee of the Radio Manufacturers Association agreed that specific standards would have to wait for further research, but the members did agree on a set of goals for the industry that included the development of "quiet and satisfactorily illuminated picture equipment for the home" at a reasonable price.[23]

By the following year, the television committee of the manufacturers association had new members primarily interested in all-electronic television. Thereafter, the FCC increasingly played an active role, encouraging the industry to develop industry-wide standards. Late in 1935, after visiting RCA, Philco, and Farnsworth laboratories to observe their television systems, chief engineer Tunis A. M. Craven and the other commissioners concluded that the industry needed to conduct more research before they could authorize commercial service.[24]

During crucial hearings held by the FCC in June 1936, industry and government officials explored a number of issues in preparation for an international telecommunications conference in Cairo, Egypt. They discussed the use of the entire frequency spectrum in the United States, with special emphasis on the future allocation of frequencies higher than 30,000 kHz for television, FM radio, facsimile, and other new services. The development of new tubes and transmitters helped open up these new bands. *Broadcasting* described the hearings as the "most important before a government agency since the birth of broadcasting fifteen years ago." It reported that "some 600 representatives of the best brains and the biggest capital in radio" were present "to listen to the views, experiences, and demands of about seventy-five leaders from all branches of the radio arts, industry, and science." Prior to the June hearings, the same publication argued that it seemed "a foregone conclusion that visual broadcasting . . . will hold the greatest interest" at the hearings.[25]

As with the earlier policy debates about the future of radio broadcasting, the tension between technocratic and nontechnocratic decision making played an important role in the 1936 debates about the future requirements for television. However, the political climate of the New Deal Roosevelt administration provided a new context for resolution of crucial issues. The man Roosevelt had appointed to head the FCC emphasized not only cooperation between government and industry but also economic and social planning "to foresee trends and the possible effect upon existing industries" of new technical developments. Whereas the FRC had largely been guided by technocratic values and laissez-faire assumptions about the relationship between technology and society, during much of Roosevelt's administration the FCC did not try to limit its work to technical problems but emphasized interrelated economic, social, and political concerns. Even a representative of the commission's technical side—Craven, the chief engineer—argued that "we can't be guided solely by scientific factors if they are in conflict with social and economic factors."[26]

When referring to "engineering hearings," official statements explicitly emphasized that the FCC wanted to take into account not just the technical qualifications of proposals but also their "relative social and economic importance." Thus Craven stressed that the 1936 engineering hearings would evaluate the "economic factors of visual broadcasting, as well as the possible effect on other broadcast services and upon other industries such as newspapers and motion pictures." Instead of exclusively relying on technical experts in order to avoid political involvement, members of the commission, including chief engineer Craven, met publicly with President Roosevelt and other "high administration officials" to discuss the hearings. Further, to emphasize the inclusiveness of its deliberations, the FCC reported that it had solicited views at the engineering hearings from "all persons and organizations interested in the development of the radio art," from manufacturers and broadcasters to labor organizations and education groups.[27]

Much of the testimony about television at the engineering hearings held by the FCC reflected the commission's effort to take into account, explicitly, social and economic concerns. Groups opposed to immediate authorization by the commission of television standards specifically asked permission to "venture beyond technical considerations." They insisted that the commission deal with "social and economic principles which must underline any policy of future development" of television. Most of the independent radio manufacturers wanted the commission to resist freezing television standards on the grounds of RCA's "monopolistic practices in patent pooling." Eleven manufacturers, including Philco, Crosley, Stromberg-Carlson, and Motorola, argued that the "Radio Corporation of America, by reason of the pooling of relevant patents of vir-

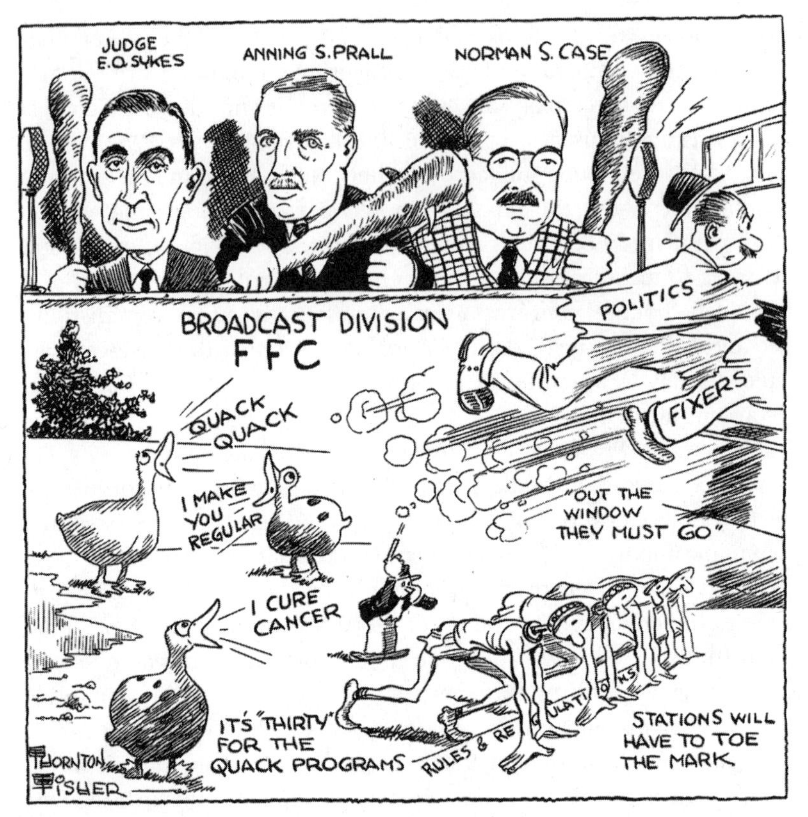

Even before James Lawrence Fly became chairman, the Federal Communications Commission reflected New Deal efforts to make sure broadcasters supported the public interest. The commission threatened to crack down on advertisements for quack medicines and related programming "abuses," but First Amendment concerns always made these efforts problematic. Note that the cartoonist wrongly used FFC for FCC. From *Broadcasting* (April 1, 1936): 7.

tually the entire electric industry, is in control of broadcast transmission and the manufacture of radio broadcast receivers, and one question to be considered is how far that control will be allowed to extend into the television field." These independent manufacturers, along with CBS, RCA's main rival in broadcasting, warned that freezing standards would not only help RCA and its subsidiary NBC gain economic control of television but potentially might also help create a monopoly in ideas. Participants at the engineering hearings thus insisted that the technical issue of authorizing commercial television standards raised "highly fundamental considerations such as the safeguarding of the free flow of ideas and information which is the cornerstone of American Democracy." Groups representing the motion-picture industry also insisted that the "commission take cognizance of the economic and cultural problems" of television. Robert Robins of Cinema Sound Corporation testified that "television should be kept free from advertizing sponsorship . . . and must be offered to the recipients on a service charge basis."[28]

As a result of the 1936 engineering hearings, the FCC decided to continue to refuse to approve commercial television standards. Testimony at the hearings had underscored the existence of important disagreements within the industry. Although RCA executives opposed a slow pace for television standardization, they did not argue for quick action by the commission. Instead, RCA tried to avoid controversy at the hearings by following a strategy of technocratic legitimation, arguing that it consistently made recommendations based on "engineering analysis and adequate experimentation." Farnsworth Television, which was working on developing its own system, argued more strongly for authorization of commercial standards. The company representative asked rhetorically if the commission wanted to "wait until somebody pulls a new invention out of the blue, or shall we give . . . the public what it is now demanding in a way which we now can do." Although it rejected this argument, the commission did encourage further experimentation and actively worked to cooperate with the Radio Manufacturers Association to get the industry to arrive at a consensus about standards.[29]

Throughout the next two years, the manufacturers association continued to deliberate on new standards for television to submit to the FCC for approval. Industry disagreements delayed action immediately following the engineering hearings, but the association did arrive at a "five-point plan" to guide further discussion: one set of television standards for the country; high-definition images "approaching home movies in clarity"; "as near . . . as possible" to nationwide service coverage; "simultaneous broadcasting of more than one program in as many localities as possible"; and receivers as inexpensive as possible.[30] Significantly, the guidelines did not attempt to limit the association to consideration of technical issues but explicitly included social and economic concerns.

However, when the committee on television of the Radio Manufacturers Association met in September 1938 to finalize a set of standards to submit to the commission, it decided to restrict its deliberations to technical criteria. W. R. G. Baker, the director of engineering of the association and a GE employee, had sent a letter to the committee asking if the members might want to consider how different companies would benefit from owning patents required for apparatus designed according to the standards under consideration. After discussing the proposal, the committee reported to Baker that "since it was divided in opinion as to the usefulness of such lists as might be compiled, the committee did not take definite action on this matter." But in its official report to the FCC explaining "the reasons for the basic decisions made in the formulation of the Radio Manufacturers Association Television Standards," the committee stated emphatically that it "only had but one objective in mind; the best technical solution." According to the representatives of the trade association, "the matter of patents was not considered as all those active in the work felt that only the most practical and technically sound standards would be adequate for the United States."[31] The association thus avoided conflict and achieved closure in its deliberations by trying to limit its work to technical decision making.

Although, as late as 1937, the FCC continued routinely to reject industry demands for commercial broadcasting, during the following year industry pressure for commercialization became harder to ignore, especially after September, when the Radio Manufacturers Association submitted seventeen recommended standards, including a frame frequency of 30 complete pictures per second with 441 scanning lines per frame. Around the same time, the *Milwaukee Journal,* which had been operating an experimental television station pending a change in commission policy, applied directly to the FCC for a commercial license to operate a system based on "RCA specifications for television equipment." Not surprisingly, RCA's David Sarnoff, who planned to begin public broadcasting and the selling of receivers in time for the 1939 World's Fair, supported the Milwaukee application. As we have seen, although Sarnoff's decision was calculated to put maximum pressure on the commission, he was also responding to pressure from competing manufacturers—specifically Communicating Systems, Inc., and DuMont. These smaller companies were threatening RCA's market position by selling receivers to consumers so they could observe experimental broadcasts. Sarnoff also believed his company needed to begin regular broadcasting in order to preempt other telecasters—including DuMont, Philco, Don Lee, and CBS—whom he feared were ready to push alternative systems onto the public.[32]

Commission Response during 1939

The FCC refused to respond to any of the various attempts in 1938 to convince it to approve standards or authorize commercialization. Sensing that important differences still existed in the industry, the commission decided to form its own television committee, in January 1939, to investigate the issues more fully. During April, the members of the committee—commissioners Craven (the former chief engineer), Thad. H. Brown, and Norman S. Case—"visited and conferred" with companies developing television, including well-established firms like Farnsworth, Philco, GE, RCA, CBS, and DuMont as well as small upstarts like the National Television Corporation and the International Television Radio Corporation. Discussions with these television interests confirmed the commission's suspicions that the industry was not unanimous in its presentation of standards. The committee concluded that "there are two divergent schools of thought." One group argued "from a technical standpoint" that television was ready for commercialization using the standards proposed by the Radio Manufacturers Association; another group was convinced "that the proposed standards are not sufficiently flexible to permit certain future technical improvements without unduly jeopardizing the initial investment of the public in receivers." In making a decision about the future of television, the committee thus faced a fundamental tension: wanting to "encourage American inventive genius" by allowing private industry to receive a financial return on its investment, while at the same time taking into account "the interests of the public," which the commission did not think would benefit from premature obsolescence of receivers.[33]

The television committee presented the commission with its first official report in May, just in time for the demonstration by RCA and other broadcasters and manufacturers of television at the World's Fair. The fifteen-page document responded primarily to the recommendations of the manufacturers association for technical standards. On the one hand, the committee made a technocratic contention that "the technical ingenuity of American inventive genius must solve the problem and indicate the road television development ultimately will follow." According to the commissioners, "it is imperative . . . that our Government take no action which retards logical progress in the further development of television." On the other hand, the committee acknowledged that "careful coordinated planning is essential not only by various elements of the industry, but also between the industry as a whole and the Federal Communications Commission." Instead of limiting its discussion to technical issues, the commissioners explored social and economic concerns that they thought were "inherent in the consideration of any technical standards." Especially important was the issue of flexibility. The committee believed technical standards needed to be sufficiently flexible so that when the industry introduced new transmitters, consumers could still use their old receivers. Standards should be able to adapt to new innovations to "permit not only improvements in quality but also radical reduction in price." Unlike the committees of the manufacturers association, the commission's television committee included "the question of patents" as a crucial consideration in the evaluation of technical standards. The commissioners did not want to make decisions that would "favor unnecessarily the patents of one person over those of another," and they argued that standards "which insure broad patent bases are preferable to those which narrow the base to a few patents."[34]

The committee's final recommendation in its first report was to "neither approve nor disapprove the standards proposed by the Radio Manufacturers Association." The report emphasized that the public should not interpret this decision as implying that the commission "believes the proposed standards to be objectionable." But their evaluation also emphasized that they did not think the standards proposed by the manufacturers association "contained a maximum degree of flexibility." Nor were they convinced that receivers built according to these standards would be cheap enough to create a mass market for television by tempting the average consumer. Although the final decision to take no action seemed to indicate that the commission was swayed by a commitment to technocratic principles, the decision-making procedures outlined in the report clearly indicated that the commission was prepared also to seriously consider social and economic issues as part of its commitment to the public interest.[35]

The May report left many in the industry dissatisfied, especially RCA and some of the other companies that tried to use the World's Fair to gain support for commercial operations. Historians of technology have written extensively on the tendency of "technological enthusiasts" to present new technologies in utopian terms. The corporate promoters of television should be viewed as exemplars of this tradition. Sarnoff predicted that television would lead to the "unification of the life of the nation, and at the same time the greater development of the life of the individual." He believed that education would be reinvigorated when "the best teachers in the land" began to teach to "millions of children simultaneously." According to Sarnoff, the spiritual life of the country would also "rise to new spiritual levels." "With television the listeners can participate most intimately in the services of the greatest cathedrals . . . and observe directly the solemn ceremonies at the altar." Rather than replace the theater and disrupt similar established institutions, Sarnoff was convinced that the "rising cultural level" accompanying television would lead to a "rebirth of local community theaters for the production of legitimate drama, musical performances, dances, and the like." Similarly, the media analyst Orrin E. Dunlap Jr. argued that television would help raise politics to a new level of integrity. When political candidates began to use television, he predicted, "sincerity of the tongue and facial expression [would] gain in importance. . . . The sly, flamboyant or leather-lunged spellbinder has no place on the air. Sincerity, dignity, friendliness and clear speech . . . are the secrets of a winning telecast."[36]

The FCC's television committee responded to the industry pressure accompanying these and other utopian predictions by releasing a second report in November. This report focused especially on the call from a number of broadcasters for a relaxation of the commission's rules restricting commercial sponsorship for television programs. Although industry officials had expected to sell more than twenty thousand television sets in the greater New York City region by the end of 1939, they were frustrated by actual sales of fewer than one thousand. With only three stations broadcasting to the public at the time of the second report, the potential viewing audience was necessarily limited. Industry leaders hoped that removal of rules against commercial operations would stimulate public interest in television by providing much needed advertising revenue to support high-quality programming, new stations in new markets, and networking facilities to connect broadcasters. Manufacturers and broadcasters had already incurred expenses of more than $15 million in developing electronic television (two-thirds of this by RCA alone) and many were anxious to begin to receive a return on their investment.[37]

Although the committee expressed sympathy with the industry's economic concerns, it concluded that "the claimed advantages of removing the restrictions against commercialization of television do not outweigh the potential disadvantages." The commissioners still worried that "premature commercialization" might "retard logical development." Specifically, they were concerned that a large number of "unfitted applicants" interested in short-term economic gain rather than the public interest would quickly saturate the market with low-quality broadcasts, resulting in a public backlash.[38]

The committee did, however, try to "suggest a clarification and simplification of existing rules," by emphasizing that "sponsorship is not prohibited, provided such sponsorship and the program facilities or funds contributed by sponsors are primarily for the purpose of experimental program development." Further, despite not wanting to freeze standards prematurely, the committee also seemed to modify its earlier position, undoubtedly in response to industry pressure, by recommending that the FCC encourage new stations broadcasting to the public to use the technical standards proposed by the Radio Manufacturers Association. The report played down the industry disagreements that the commission had earlier acknowledged by arguing that systems based on these standards were now being used by "all" stations broadcasting to the public and most manufacturers selling receivers. "While the future may require changes in the Radio Manufacturers Association standards by reason of improved technical progress," according to the committee, "for the time being these standards must be used for scheduled program service."[39] The economic importance of standardization seemed clear to manufacturers and broadcasters fighting for commercialization; until the entire industry adopted universal standards, potential consumers would remain hesitant to purchase expensive receivers that might soon become obsolete.

The commission adopted, "with minor modifications," the recommendations of the second report of the television committee in December and announced that it would hold hearings beginning January 15 to give interested parties an opportunity to express their views. But testimony at the hearings, which lasted eight days, confirmed the commission's earlier findings that serious divisions existed in the industry about adoption of technical standards and authorization of commercial broadcasting. RCA and Farnsworth, the two companies that most observers believed held the major patents to the technical system implied by the standards of the manufacturers association, expressed strong support. However, three other major companies—Zenith, Philco, and DuMont—expressed significant misgivings. Zenith, which had delayed committing resources to television research, argued against freezing standards prematurely and called for further research before the FCC took any action. Philco and DuMont wanted flexible standards that would help avoid the problem of obsolescence and allow the companies to market alternative systems. DuMont engineers had not taken part in the committee of the manufacturers association that developed television standards; Philco engineers had been active participants and had originally agreed to the standards at the meetings, but now believed the industry needed to reconsider the earlier findings in order to take into account new developments. Instead of fixing the frame frequency at 30 frames per second and the line frequency at 441 lines per frame, their alternative proposals called for standards that would accept a range of options. DuMont proposed a system that would accept anywhere from 15 to 30 frames per second and from 441 to 800 lines per frame. Philco proposed an additional six months of research to consider fixing the frame frequency at 24 frames per second (fps) while allowing any number of lines more than five hundred lines per frame (lpf).[40]

DuMont and Philco both believed their alternatives would give higher-quality pictures than the manufacturers association's recommendation of 441 lpf / 30 fps. Engineering disagreement on this question resulted from the existence of a complex trade-off among a number of different factors, including the subjective element of judging acceptable picture quality. As in the case of the regulation of the radio spectrum, a final determination involved something more than rule-governed technical criteria. Experimenters could obtain a better picture by increasing the number of scanning lines per frame, but unless they reduced the frame frequency the television signal would require an increase in channel width. Due to the demands of other services for a limited number of frequencies available in the radio spectrum, the FCC did not want to increase the established channel width. Commissioners told industry representatives that they should "explore all possibilities of improving performance within the present 6000 kilocycle band assigned."[41]

Given this channel limitation, the only way to increase the number of lines was to lower the frame frequency. But if experimenters lowered the frame frequency too much, the picture appeared to flicker. DuMont engineers believed they could solve this problem through the use of a special "retentative screen" they were developing that enhanced the "persistence of vision" of the eye so that no flicker occurred even with the frequency set at 15 fps.[42] The company also contended that, unlike its proposed system, the standards recommended by the manufacturers association lacked the flexibility to produce high-quality pictures on large, twenty-inch, screens.

In its testimony, Philco pointed out that the British were using a frequency standard (25 fps) lower than the trade association's recommendation. According to Philco, there was "virtually unanimous agreement among the technical experts" that "were it not for the question of 'hum' raised by the usual 60-cycle power supply in most localities," its 24 fps proposal "would be preferable" to the 30 fps recommendation of the manufacturers association.[43] The British did not have to worry about this problem because they used a higher power-supply frequency. Both Philco and DuMont argued that they could develop circuits to eliminate the potential 60-cycle interference in the United States. Thus, both companies believed the Radio Manufacturers Association should reconsider its recommendations in order to give the industry time to develop flexible standards based on new research.

Other aspects of these standards were also in dispute, most notably the proposals for synchronization and polarization. The synchronization standard, which was essential for the proper functioning of the lock-and-key system, established the character of the transmission pulse that synchronized the picture signal from the transmitter with the signal creating the image in the receiver. DuMont favored a pulse with the flexibility to adjust the receiver to different transmitted frame frequencies automatically. Philco argued that vertical polarization was superior to the recommendation of the manufacturers association for horizontal polarization. Polarization determined the structure of the antenna and the nature of the transmission wave. Vertical polarization had many advantages, according to Philco engineers, including better ability to discriminate against noise, greater efficiency in using transmissions in areas where the signal was weak, and less expense for consumers. They also pointed out that most other countries had adopted vertical polarization for sound broadcasting and that the United Kingdom had already adopted the standard for television service.[44] Thus, industry disagreement turned on issues linked to technical flexibility and uncertainty. A final decision would depend on trade-offs among a number of different factors.

Limited Commercial Operations Authorized

On February 29, 1940, the FCC adopted new rules based on the findings of its television committee and the industry testimony at the January hearings. The commission decided to create two classes of stations "in view of the fact that certain groups in the industry wish to stress technical research and others program experimentation." Beginning on September 1, the government would allow "limited commercial operations" on the stations that were committed to programming experimentation, designated "Class II stations." But the commission's unanimous decision stressed that "emphasis on the commercial aspects of the operation at the expense of program research is to be avoided." The rules also required that stations use sponsors only to pay for program production, not to cover costs connected with the use of broadcasting facilities.[45] The commission took this compromise position in response to widespread industry demands for relaxation of the earlier rules barring all commercial operations. Some companies, most notably DuMont, rejected standardization, but they joined RCA and Farnsworth in requesting commercial authorization.

Although in its earlier decisions the FCC had not made a sharp distinction between commercialization and standardization, the new rules signaled a change in policy. The February report emphasized that broadcasters and manufacturers should not interpret authorization of limited commercial operations as permission to adopt technical standards for commercial service. Based on industry testimony and demonstrations, the commission was convinced of the importance of avoiding freezing standards because of "a substantial possibility that the art may be on the threshold of significant advances." The commission specifically warned individual companies not to attempt to impose standards on the industry. Based on testimony at the hearings, the commissioners believed that even RCA and Farnsworth were not ready to freeze standards in such a way that further research would be impossible. RCA counsel F. W. Wozencraft had claimed that "as far as RCA is concerned, we don't ask that the standards be frozen or that the commission approve the standards in a way which will make it more difficult for anyone else who has other standards at any time." In order to authorize limited commercialization while at the same time avoiding freezing standards, the commission encouraged the industry to pursue research aimed at developing the kind of flexible standards proposed by DuMont and Philco.[46]

If the February report seemed to side more strongly with DuMont and Philco than with RCA and Farnsworth, this had been even more apparent during the January hearings, when the commissioners harshly questioned representatives of the Radio Manufacturers Association about possible dominance by RCA. They probed why the recommended standards did not represent the unanimous judgment of the industry and, especially significant for this discussion, raised important questions about the legitimacy of using the engineering department of the manufacturers association as a source of technical advice for decision making on policy.

Much of this severe questioning of the industry came from James Lawrence Fly, formerly the general counsel for the Tennessee Valley Authority and an especially strong supporter of New Deal activism; President Roosevelt had appointed him chairman of the FCC in fall 1939. Probably Fly's most important work on the commission was to complete an investigation, first initiated in 1938, of monopoly control by network radio. The resulting "Report on Chain Broadcasting," issued in 1941, became a blueprint for the partial reform of the radio industry. The U.S. Supreme Court upheld Fly's policies in 1943, and, among the most notable results, NBC was forced to sell one of its two networks, which became the basis for a new entity, the American Broadcasting Company (ABC). Not surprisingly, then, at the January 1940 hearings, Fly particularly wanted to find out "to what extent a single company group has more than one membership on any of the committees" of the Radio Manufacturers Association. The association lost much credibility when representatives were forced to acknowledge, after first giving evasive answers, that a number of committees or subcommittees each had at least two members from either RCA, RCA Manufacturing Company, NBC, or RCA Licensing Company. Further, no other company seemed to have similar multiple representation.[47]

Fly also closely questioned the representatives of the trade association about the process used to develop the standards recommended to the commission. He was especially interested in discovering if the decisions represented "a composite judgment or, in effect, a compromise of judgments." Fly found particularly troubling the fact that Philco and Zenith had voted against certain elements of the standards and that DuMont had not taken part in the work of the association's committee. The manufacturers association lost further legitimacy when its representatives admitted—without thoughtful elaboration on the implications for policy making—that engineers' decisions had involved uncertainty and technical flexibility. Baker testified that "no set of standards are ever a finished product"; they merely represent "the best intelligence of the industry at the time" they were formed. The idea that engineers had to vote and pass technical decisions based on agreement of some percentage representing the majority seemed to indicate that nontechnical considerations were playing a role in the decision-making process. If this was the case, Fly wanted to know, then why had the engineers not taken into consideration the question of patent control. Baker emphasized that "the engineers, to my knowledge, stayed entirely clear of any patent considerations and interested themselves in the formulation of a better standard, which would give the best service to the public."[48]

The inability of engineers representing the manufacturers association, in the view of the commission, adequately to deal with the relationship between technical and nontechnical factors in the formulation of television standards played an important role in the commission's "loss of confidence . . . in the integrity of the procedure of the engineering department." Although the FCC, at least since the mid 1930s, had generally emphasized the importance of taking into consideration the complex interrelationship between these two kinds of criteria, the trade association, especially the members who supported the standards at the commission hearings, consistently tried to maintain a sharp boundary and to legitimate decisions through exclusive technical arguments. Engineers testifying at the January hearings as representatives of the manufacturers association specifically limited their testimony to "technical procedure." They thought the commercial television committee of the manufacturers association should deal separately with "commercial" considerations; however, by drawing a sharp boundary between two different kinds of considerations and at the same time somewhat inconsistently treating the development of standards as basically an engineering problem that engineers would solve, they only helped provide more reasons for the commission to question the integrity of the Radio Manufacturers Association. Industry leaders like David Sarnoff supported the engineers' technocratic views. At the association's board of directors meeting immediately following the commission hearings, he complained that "the radio law never contemplated that the Federal Communications Commission should have anything to do with patents or merchandizing policies, or rates, or prices at which sets are sold." When it comes to evaluating technical issues such as the development of television standards, "we must," he insisted, "in our operating positions be guided by our engineers' recommendations."[49]

The commission's questioning of this technocratic commitment by the manufacturers association and its loss of faith in association engineers led Fly to announce at the close of the January hearings that the FCC was considering ignoring the association and appointing its own committee of engineers from "leading manufacturers" who would unanimously recommend television standards for the country. The commission did not follow through with this threat in its February report, but further controversy during spring 1940 seemed to provide additional evidence that the industry could not reach agreement without government intervention. During a meeting of the subcommittee on television standards of the manufacturers association on February 29, the same day as the commission report, a Philco engineer who chaired the committee resigned to protest what he believed were efforts by association members, especially representatives of RCA, to freeze standards. Significantly, he also raised the possibility of having "a Government Television Bureau established with adequate appropriations to study and set up various television transmission systems and establish proper standards." This event had recently occurred in Britain, where a government commission had decided between alternative technical systems.[50]

The commission took more seriously Philco's accusation that RCA was attempting to freeze television standards when, beginning on March 20, Sarnoff's company launched an intensive promotional campaign to convince the public to purchase television receivers. The commission was particularly upset because RCA advertising failed to acknowledge the experimental and limited basis of television broadcasting. Specifically, the advertisements did not reveal that RCA stations were broadcasting only for about two hours a day and that RCA receivers would not be able to receive programming from other transmitters because they were using different technical systems. The commission feared that RCA's actions might saturate the consumer market with expensive receivers, which would result in a curtailment of further research and a crystallization of standards to the proposal of the manufacturers association—actions the commission had specifically warned the industry to avoid in its February report.[51]

Limited Commercialization Suspended

With these new concerns, the FCC announced on March 22 that it would hold further hearings in April to "determine whether research and experimentation and the achievement of higher standards for television transmission are being unduly retarded by the action of the Radio Corporation of America or its subsidiaries, or any other licensees." Commissioners emphasized that they were not explicitly attempting to regulate advertising or trade practices but to exercise a clear congressional mandate, under the 1934 Communications Act, to promote the development of the best transmission standards consistent with the public interest. The commission effectively suspended its previous authorization of limited commercial operations to begin on September 1 until after the conclusion of the hearings.[52]

The public reaction to this decision was by no means uniformly favorable. Fueled partly by the utopian predictions of Sarnoff and other industry officials, many individuals had high expectations of television's imminent development into a powerful new industry that would help pull the country out of the Great Depression. They feared that "a big industry has been struck down, with the inevitable result that more unemployment will follow." Many of the commentators upset with the FCC's decision were also longtime critics of the New Deal; they viewed the commission's action as another example of "the wave of arbitrariness and usurpation of power which has swept Washington lately." The *New York Herald Tribune* used this as an occasion to lash out against "bureaucrats" who "have an irrepressible desire to extend their personal influence and power and reach out to control much more than was originally intrusted to them." The conservative critics of the commission found support for their attacks on big government in the technocratic argument, already expressed by Sarnoff and especially by members of the radio commission before him, that "public interest, convenience, and necessity . . . were never intended by Congress to cover anything but" the "purely technical" problems involved in allocating wavelengths. Their position seemed strengthened by the harsh criticism leveled at the commission's decision by Commissioner Craven, whom David Lawrence of the *Washington Star,* a strong critic of the New Deal, contended was "the only man on the commission who knows anything about technical problems." Craven characterized Fly's fear that RCA's commercial activities would help freeze standards as "absurd." He contended that "nothing can stop scientific research and technical progress in a free democracy if incentive is not discouraged by Government." But Craven did not argue that technical progress was simply driven by an internal logic of development existing in a social vacuum. Rather, in a free, market-driven consumer democracy, the public was "the wisest judge of scientific achievement and will be most effective in securing the technical improvements if desired." Craven feared that the FCC, by interfering in this process, might "create such confusion as to retard the development of television and discourage the incentive and initiative of private enterprise."[53]

The arguments of these opponents of FCC policy reflect tensions and potential contradictions in the work of regulatory commissions. As we have seen, critics wanted the commission to follow the advice of the technical experts. "It seems little short of presumptuous on the part of the Federal Communications Commission to oppose its own views to those of experts," said the *New York Times*. But the main technical expert on the commission thought consumers should play the most important role in determining technology policy. Further, even Craven was not entirely consistent in his emphasis on a free democracy of consumers. The votes of some consumers apparently counted more than others. According to Craven, wealthy individuals would have a stronger voice in determining technological development: "The burden of experiment falls on wealthy people, as it should, to pave the way for ultimate inexpensive television to all."[54]

Members of Congress sympathetic with the concerns of the opponents of the commission called for an investigation of FCC actions during the spring of 1940. In response to a resolution submitted by Senator Lundeen of Minnesota, the Senate Committee on Interstate Commerce held hearings in April, concurrently with the commission's own hearings, "to ascertain whether the commission has exceeded its authority, and whether it has interfered with the freedom of public and private enterprise." With the involvement of Congress, the political and social aspects of the technical controversy became more prominent. As *Business Week* pointed out, "a first class political football was in the making."[55]

An analysis of the testimony at the Senate and commission hearings in April illuminates the complex tensions already apparent in the earlier public statements of Craven and other opponents of FCC policy. Most participants at these hearings, despite their differences, tried to legitimate their positions based on engineering testimony. Sarnoff justified RCA's involvement with the manufacturers association and support for its standards by emphasizing the overwhelming technical expertise of the company and its preeminent role in developing television. He reminded the Senate committee that "hundreds of [RCA] engineers have been engaged in developing television, and the current rate of expenditure for this work by the RCA alone is about 2,000,000 dollars annually." To demonstrate how his company had "done more to develop high television standards than any other organization in the United States," he pointed out the numerous technical reports produced by RCA, including "229 papers and reports to scientific societies, 671 additional technical reports, and 2 major textbooks." Fly, arguing along similar lines but not to make the same point, said he wanted to base national policy on "preponderating engineering testimony." The decision against approving standards, he claimed in a radio address on April 2, was "merely a reflection of the engineering opinion in the television industry." Philco, rather than allow the "nonexpert public" to "participate in the technical process of choosing between the comparative merits of two or more systems," wanted to leave it to "qualified experts in the commission's staff and the industry." Laboratory experimentation was thus preferable to public experimentation.[56]

Despite apparent agreement among many of the participants in the hearings about the important role of technical experts in legitimating public policy for television, subtle disagreements became prominent, including different views about the proper relationship between technical and nontechnical decision making. FCC chairman Fly said he thought technical evaluation by engineers should also take into account interrelated economic and social concerns, including especially the patent policies of different companies. As we have seen, Sarnoff generally wanted to draw a sharp boundary and let narrowly defined technical criteria dictate policy, but in his Senate testimony he somewhat inconsistently criticized Fly for overemphasizing the importance of the technical factor of the lock-and-key relationship. "One gets the impression from a good deal of confused discussion on the subject," he argued, "that if the key is changed or if the lock on the house is changed, you have to burn down the house or otherwise dispose of it." According to Sarnoff, the FCC also needed to take into account economic calculations when evaluating the engineering problem of standardization, including his estimate that a $400 receiver could be adapted to fit the requirements of a new technical system at a cost of no more than $40.[57]

Engineer Lee de Forest, who remained active in electronics experimentation after his early contributions to radio engineering and continued to call for the social responsibility of engineers, expressed, in a written brief to the commission, a much more extreme view about the social and economic role of engineering evaluation. De Forest warned that "under no circumstances must we allow the mere technicalities of this industry to confuse the greater issues at stake." "Social factors" were "more fundamental and of more lasting importance." Engineers and inventors, he reminded the commission, were "also Americans with opinions concerning the commercialization of television and its broad possible effects upon the future of the United States." Specifically, de Forest thought engineers should put aside their company's interests and take into account the "monopoly issue" and patent policy. "Our Constitutional rights to freedom are greater than patent rights," de Forest stated, "and even though the television industry might come to employ a million people at this time when we sorely need such a business if it should mean giving up our Republic or the passing on of a tyranny to our children, we would be better off never to have television."[58]

Although the conflicting testimony at the Senate and commission hearings in April hinted at the possibility of a major political controversy involving the FCC, Fly managed to maintain the upper hand. Congress took no further action at the conclusion of the Senate hearings, and President Roosevelt expressed strong support for the commission's decision to postpone authorizing commercial television broadcasting in order to prevent the development of a monopoly.[59] The commission's May 28 report on the April hearings reaffirmed its earlier evaluation that "the industry as a whole does not share the RCA view of forging ahead." It suspended the February order allowing partial commercialization until the industry could come to an agreement on standards.

The FCC again expected engineers to play the most important role in this process. "The commission will consider the authorization of full commercialization," the May 28 report concluded, "as soon as the engineering opinion of the industry is prepared to approve any one of the competing systems of broadcasting as the standard system." But the report also emphasized that the determination of technical standards should take into account economic and social concerns, especially the problem of patent control and the avoidance of monopoly. According to the commission, "the positions of the different companies on this whole problem cannot be viewed with total disregard of the patent interests of competing manufacturers which find expression in a desire to lock the scientific levels of the art down to a single uniform system based in whole or in part upon such patents." The commission wanted to "guide" the industry away from potential "monopoly" and toward "healthy progressive competition." Thus, the FCC not only affirmed the central role of engineers in policy making for television but also reemphasized the interrelationship between the "technical" problem of standardization and the "socioeconomic" problem of commercialization.[60]

Establishment of the National Television System Committee

Although the commission wanted industry engineers to play a central role, it was still critical of the record of the various standards committees of the Radio Manufacturers Association. The commission now felt prepared to follow through on its earlier threat to establish a new technical organization to advise government regulators about standards. In June, Commissioner Fly and the commission's chief engineer met with two officials from the manufacturers association—the president and the head of the engineering department—to discuss the organization of a new group, the National Television System Committee (NTSC). Fly was mainly concerned that this new organization should represent all sectors of the television industry. The manufacturers association was still to play an important role by "sponsoring" the new organization and appointing the members, but Fly specifically set up the committee to be "independent of any other organization." No one company, most importantly RCA, would have more than one representative on each of the panels or subcommittees established to investigate individual problems. Further, the system committee would not limit membership to organizations belonging to the manufacturers association. The committee thus included DuMont, the important company that had not participated in the standards committees of the association. Altogether, fifteen organizations appointed members to the central board of the system committee, mainly major manufacturers and broadcasters "enjoying a national reputation in their work on electronic methods of television." There were also two nonprofit groups, the National Association of Broadcasters and the Institute of Radio Engineers. Although the central board did not include as regular members representatives of smaller "interested companies" without a national reputation, 169 individuals representing a wide variety of interests took part in the deliberations of the committee's different panels. These participants included not only industry engineers but also independent consulting engineers and technical experts from leading colleges and research laboratories. FCC engineers did not participate directly in the work of the system committee, but they maintained close contacts. In July, FCC engineer Jett assured the new group that "the engineering department of the commission is anxious to cooperate closely with the committee and its technical subcommittees so that advantage may be taken of our various points of view." During the following months, members of the engineering department helped organize demonstrations of different television systems and suggested research and experimentation for the system committee to perform. With the establishment of the new technical advisory organization, Commissioner Craven retracted his earlier criticisms and supported the suspension of commercialization until "the scientists of the industry can come to an unbiased agreement."[61]

After an organizational meeting of the system committee in July, which FCC officials also attended, the nine panels began meeting in September. Baker emphasized the importance of this work by reminding members that television "probably [will] have more effect on the life of the American people than any system known today." Seven of the panels of engineers dealt with specific technical problems, including synchronization (panel eight), polarization of the signal wave (panel nine), and picture resolution, or the consideration of frame frequency, line density, and other factors determining picture detail (panel seven). The first two panels explored more general problems of television standardization. Panel one evaluated all operating and planned television systems, both foreign and American. Panel two considered "the influence of physiological and psychological factors in the determination of television system characteristics." During the next five months, the panels met sixty times, observed twenty laboratory demonstrations, and wrote nearly one hundred reports.[62]

CBS added a new complication to the committee's work when it surprised the industry in August by announcing the development of a color-television system developed by its chief engineer, Peter Goldmark. Because it was not compatible with existing monochrome standards, Goldmark's "field-sequential system" offered a radical alternative. It added color to a standard monochrome system through the use of mechanical discs adapted from the mechanical era of television. The camera used a disc made up of color filters spun behind the lens. The apparatus scanned each picture frame in red, blue, and green. The system then transmitted sequentially six double-interlaced fields to make up one image in the receiver, which reproduced the picture using a synchronized rotating color disc. Because of the increase in the number of frames needed to compose one image, Goldmark increased the frame frequency to 60 per second (double the standard recommended by the manufacturers association). To compensate for this faster frame frequency, he lowered the line definition to 343 (instead of the 441 recommendation of the manufacturers association). The 343 lines were divided into odds and evens—

171.5 of each. The first frame was made up of the 171.5 odd lines in red and the 171.5 even lines in green. The second frame was composed of odd lines in blue and even lines in red. The third frame included the odd lines in green and the even lines in blue. The system scanned three frames or six fields every one-twentieth of a second to create an entire picture. Goldmark based his color system on research by John Baird and other engineers from as early as the 1920s. His major innovation was to find a way to raise the number of lines by 50 percent over earlier color systems. After the first demonstration to the commission, company officials claimed that "color, in the opinion of those who have seen it, added depth to the whole picture and eliminated the flat quality that many people have felt exists in black and white television."[63]

During the deliberations of the panels of the system committee in the fall of 1940, it became clear that three major issues divided the industry: the problem of accepting fixed or flexible standards, the proper method of synchronization, and the allowed standards for picture quality (most importantly, frame frequency and number of lines per frame). In a tentative report presented during hearings before the commission in January 1941, the committee announced that it had reached agreement on all issues except synchronization and the number of picture lines. Generally, the completed standards affirmed the previous recommendations of the manufacturers association. The committee rejected two proposals for flexibility in favor of fixed standards. Members decided that DuMont's proposal for receivers that would accept a continuous range of scanning rates would be expensive and difficult to adjust and would likely produce pictures having an unacceptable amount of flicker. The committee rejected CBS's proposal for flexible standards that would accept either monochrome or color transmissions on similar grounds. The committee feared that the mechanical element in the color system would make the receivers too expensive and reintroduce problems bypassed by the new electronic era. Further, CBS's color system had not yet successfully demonstrated pictures reproduced from live action, only from motion-picture film. But the committee experts did make recommendations encouraging further research for the eventual introduction of color television using all the same standards except the number of picture lines and frame frequency.[64] This compromise position avoided a serious delay in authorizing monochrome television, but the potential for conflict remained (see chapter 6 for the debate over color standards).

Following the preliminary commission hearings in January, the system committee continued discussions on the final two standards, eventually reaching an agreement in March. Instead of deciding in favor of one method of synchronization, the committee recommended that since commercial receivers were already compatible with the three major proposals (supported by RCA, DuMont, and Philco), it should permit all three techniques. Agreeing on an acceptable line standard was more difficult. During deliberations in fall 1940, five of twenty-one members of panel seven had voted against the standard of 441 scanning lines per frame. New studies after the January hearing and an important demonstration by the Bell Telephone Laboratories on March 7 convinced engineers of the system committee to accept a new proposal of 525 lpf. But the new studies also seemed to indicate that any differences between the two proposals (441 vs. 525) were minor and that much room for flexibility remained. Donald G. Fink, an engineering expert whom the committee had asked to prepare a special study of the problem, concluded that the experiments at Bell Labs had shown that "in effect, any number of lines, within the range then proposed would suffice." When the main committee voted on March 8, at least one engineer, Adrian Murphy of CBS, admitted that since there was "little basis for choice as between the two discussed values, . . . he, therefore, preferred to cast his vote with the majority." Ernst F. W. Alexanderson of GE informed the same committee that "on a purely technical basis there is no marked, definite advantage for either 441 or 525 lines but that he believes that, commercially, 525 lines per frame period is slightly better than 441 lines." In this final vote, two engineers (D. E. Harnett of Hazeltine Corporation and George R. Town of the Stromberg-Carlson television manufacturing corporation) voted against changing the recommendation from 441 to 525 (at 30 fps); eleven voted in favor. However, at least two engineers who voted for the change also testified that ideally they favored another value.[65]

After holding hearings late in March, the FCC accepted the system committee's proposed standards for television. The commission praised CBS's color system but supported the committee's recommendation that before considering authorization of color standards, it should give the system a six-month trial. The commission next turned to the problem of commercialization. At the March hearings, RCA and its subsidiary network NBC expressed less interest in speedy approval for commercialization. The RCA engineering representative testified that the company "now preferred to wait until the industry as a whole was ready." Having been harshly criticized once by the commission for trying to promote commercial television, RCA was probably anxious to avoid being burned a second time. By the spring of 1941, RCA, more than its rivals, was also preoccupied with filling a growing number of defense contracts that diverted resources away from television. Ironically, companies that had opposed RCA's earlier efforts to push for commercial authorization, including Philco and Zenith, were now the strongest supporters. With the standards question settled, for the most part in line with the recommendations of the manufacturers association that RCA had strongly supported, RCA's traditional competitors anxiously sought a return on their investment through the sale of receivers to the public.[66]

On April 30, the FCC announced that it would allow commercial broadcasting to begin on July 1. However, after President Roosevelt declared a state of unlimited national emergency on May 27, the government severely curtailed television production and research. One year later, all production came to a halt. At the time of Pearl Harbor, ten television stations had already begun commercial broadcasting, including NBC, CBS, DuMont, GE, Philco, and Don Lee; only a few stations managed to continue operations through the war. But although the war delayed television, and the new technology still had to overcome other problems before it could take off, by 1941 the FCC had made the crucial decisions about standards and commercialization that provided the foundation for a powerful new industry.[67]

Tensions in NTSC Decision Making

The tendency of decision makers to apply technocratic values provides another important framework for understanding the work of the National Television System Committee in developing standards for television. As we have seen, the FCC established the committee to provide the best engineering advice free of personal or corporate bias. Public statements by officials consistently emphasized the important role of engineering evaluation in legitimating policy decisions. In his welcoming address at the organizational meeting of the system committee in July 1940, the president of the manufacturers association, J. S. Knowlson, stressed that the engineers were "not asked to attend as representatives of warring commercial interests but as scientists." He compared the birth of television to the birth of a child destined to become an "heir to an empire." Just as the child would be surrounded by different people representing "all the pulling and hauling of personal ambitions, personal interests and idle curiosity, which must always be present when a new life of world importance is about to arrive," television also was part of a great struggle for control. In both cases, according to Knowlson, leaders would avoid chaos and disorder through the intervention of technical experts. Their job was to see, "so far as their scientific skill and ability were able," that the child or the new invention had a healthy start—"not to consider personal interests of the courtiers or the princes" or "to speculate as to how this child would grow up; what sort of a ruler he might be."[68]

Members of the technical panels of the system committee also emphasized that their work provided policy making on standards with a strong basis of scientific legitimacy that would transcend industry rivalry and support the public interest. This was especially true in the case of panel two, which was responsible for studying the subjective aspects of television viewing. Although the committee seemed to be dealing with aspects of television not readily rationalized and quantified, the final report stressed that it was presenting the commission with "a pure factual document representing an assembly of what is believed to be the best scientific knowledge attainable at this time." Instead of following the example of the other panels and primarily using industry engineers, panel two relied on members who were "exclusively university and research laboratory workers," including psychologists, physiologists, and "leading experts in the fields of the theory and scientific aspects of picture scanning, of advanced radio theory, and of the photographic criteria of good picture quality." RCA engineer Goldsmith, the chair of the panel, pointed out to the commission that no member was "employed by any organization engaged commercially in television broadcasting or in television equipment manufacture and sale." He assumed that although nontechnical considerations might influence industry engineers, technical judgment would guide university and laboratory researchers to the one best standard. The panel's effort to establish a sharp boundary between technical and nontechnical considerations was enhanced by its original status as a special committee of the Institute of Radio Engineers. Before this committee was transferred to the system committee in the fall of 1940, the institute's board of directors had established a policy that its committees would consider only the purely technical aspects of standards; they wanted commercial considerations to be the responsibility of trade associations like the Radio Manufacturers Association.[69]

In the May 1941 report announcing its acceptance of the standards recommended by the national system committee and authorization of commercial broadcasting, the FCC emphasized that it based its decision on the collective judgment of the nation's best technical experts. In setting up the system committee, chairman Fly had wanted industry engineers to arrive at a unanimous decision on standards. The May report seemed to confirm Fly's success. "The standards," said the commission, "represent, with but few exceptions, the undivided engineering opinion of the industry." During the public hearing in March when the committee presented the final recommendations to the commission, officials emphasized that "technical" rather than "nontechnical" criteria had guided the engineers' decisions. Baker testified that "only the engineering phase" was considered by the committees: "The economic phase was not the province" of the system committee. One of the most important socioeconomic factors that the committee might have considered was the question of patent control. But according to Baker, the committee had "quite successfully avoided any consideration of patents."[70]

The records of the meetings of the system committee indicate that, at least ostensibly, the various panels and the main committee avoided discussing patent considerations; however, they also reveal that members did systematically take into account other "nontechnical" factors. An analysis of these issues thus helps reveal how the tensions emphasized in this chapter were resolved by the commission. The final decision brought closure to the problem of standardization and commercialization, but important tensions remained.

The system committee had a number of discussions about the legitimacy of taking into account the problem of patents, mainly under prodding from officials representing Philco. As we have seen, Philco's opposition to the standards recommended by the manufacturers association and its claim that RCA dominated the trade association had been important factors in Fly's decision to push the industry to set up a new technical committee to advise the commission. Philco believed that engineers and other officials could not easily separate decisions about standards from the position on patents of companies represented on the system committee. Even before the first organizational meeting of the committee in July 1940, Philco's president urged Baker to ask each company participating to "disclose what standards are included and what are not included in its own patent situation." Baker, however, thought that so long as the engineers tried "to do a conscientious job, the patent situation will have very little influence on the decision." He saw this as primarily a matter of morality, involving the honesty of engineers. Baker reassured Philco that the "majority of men" were "intelligent and honest" and would "sincerely try to formulate standards which will provide the best system of television."[71]

But officials at Philco did not view the problem of patents and technical decision making as primarily a matter of honesty and integrity: engineers might make decisions benefiting their companies' economic interests without being aware of it. Philco wanted to get everything out in the open so that officials could make a fully informed judgment, based on all relevant data. The company wanted the committee to get "an expression of an opinion on the part of a company that they were trying or had obtained patent coverage that would be made valuable by the adoption of that standard." Technical standards would have commercial implications, according to Philco, that might lead to "a monopoly in television through the medium of patents." The company contended that the committee engineers voting for standards and the commissioners following the engineers' advice ought "to know what they are doing and why." Philco officials also argued that the patent issue was especially relevant in cases where there existed "two equally good ways of doing a thing," one patented and one not. In these circumstances, they believed, "all would agree" that the FCC should approve the standard not tied to narrow economic interests. DuMont Laboratories, which also expressed disappointment with the final standards recommended to the commission, echoed Philco's argument about the need to view the problem of the possible socioeconomic implications of standards recommended by committee engineers as a complex structural problem rather than a simple moral issue. According to DuMont, the committee engineers' "personal integrity and competence must be undisputed," but "a man need not be dishonest to be blind to defects in the child of his own intellectual effort."[72]

Members of the system committee discussed Philco's proposal concerning patents extensively during meetings in August and September 1940. Most of the members of the main committee thought the issue would only introduce confusion and uncertainty to the deliberations. Most did not think that engineers were "sufficiently informed on the patent situations of their companies to give suitable answers." Some members even claimed that companies' patent lawyers would not be able to give definite answers because of the confused state of the field and because "a patent interest in a given topic is a rather vague thing to define or express or determine at certain stages of the work." Any company could claim that they had plans to patent research work linked to a particular standard, even if it was at a very preliminary stage. In an extreme case, the patent status of a technical development claimed by more than one company might not become completely clear until the U.S. Supreme Court made a final decision. Philco officials responded to these assertions by arguing that they simply wanted all relevant information brought "up above the table," no matter how confusing, so that a fully informed decision could be made.[73]

Members of the system committee felt justified in rejecting the arguments of these critics because they did not think the FCC had provided clear instructions requiring them to take into account the patent issue. At least one engineer believed the commission's instructions indicated that their "deliberations should be directed toward the optimum system, regardless of patents." Another member claimed that he was "not entirely sure that the instructions of the chairman of the commission were as explicit as they seem." Because members did not think the FCC had provided explicit instructions to consider patents, the system committee decided that the individual technical panels should not take any action on patents but defer the issue to the main committee. However, according to the minutes of the meetings of this coordinating committee, or board of directors, members never discussed the issue. During the FCC hearings in March 1941, after one of the commissioners had asked chairman Baker of the system committee why the main group never made a decision about the patent issue, and as Fly interrupted and cut off further discussion on the topic, the following exchange took place:

> Fly: "May I ask, Doctor [Baker], whether or not it is a fact that you conceived your job as an engineering job, to recommend the best standards that might be available regardless of patents and you didn't want to impede the work by consideration of patents with all the vagueness as the validity and scope of patent claims and that sort of thing, that you thought that any question that arose on that should come later to the commission as a matter of policy rather than to your committee as a matter of engineering."
>
> Baker: "That is true, Chairman Fly, the National Television System Committee has quite successfully avoided any consideration of patents."[74]

After having clearly expressed a nontechnocratic view, during earlier hearings and in official reports, encouraging technical experts to be sensitive to the patent implications of their decisions about television standards, why did Fly take such a strong technocratic position in this final hearing? The answer can be found if we consider the context: During spring 1941, consumers and major industry leaders anxiously awaited the introduction of television. Fly felt pressured to move the proceedings along quickly before United States involvement in the European war completely diverted the industry from television development. He testified that he wanted to avoid putting the industry "through all the expense and inconvenience of ever-recurring hearings." Fly was also preoccupied during this period overseeing the commission's "Report on Chain Broadcasting," the controversial work that resulted in a limited reform of the radio industry. This work not only antagonized the radio networks but also conservative members of Congress, who launched a series of investigations of the commission beginning in 1941.[75] Given these circumstances, Fly wanted to avoid additional controversy that might result from any action appearing to further delay television broadcasting.

Despite the strategy used by FCC and system committee officials of legitimating decisions about standards through exclusive reference to technical criteria and procedure, an analysis of the record indicates that the committee panels did consider "nontechnical" factors. When evaluating monochrome versus color standards, for example, the members of panel one decided to ask every member of the committee "their viewpoint and recommendations regarding the technical, economic, and social aspects of color television." They made a final decision against immediate authorization of color standards based on a belief that the high cost of color receivers would outweigh other potential advantages. When panel seven evaluated the different monochrome scanning systems, it reported that "with a lower number of lines, we have somewhat lower costs and it was the committee's feeling that that was of some importance." Even panel two, while publicly stressing its ability to provide purely scientific results, sought a compromise "between technical excellence and cost."[76]

The FCC was aware that the committee's panels were taking into account both "technical" and "nontechnical" criteria. Baker informed the commission during the January hearings that he had chosen members of panel one specifically based on their capacity to handle "research and development" as well as "economic problems." During the hearings in March, at least one member of panel one told the commission that he was not an engineer. Further, some engineers on the committee stressed that they felt responsible to give the FCC "the economic arguments advanced by the executives" of their companies. Other testimony underscored the large amount of uncertainty and technical flexibility in some of the standards. Goldsmith, for example, testified that aspects of picture quality studied by his panel have "no quantitative definition and no one knows quite what is the acceptable minimum."[77]

The record also indicates that both the system committee and the commission played down industry disagreement that lingered after the committee announced its recommendation for final standards. Instead of requiring unanimity, the committee only required a majority vote in order to pass standards. Although all the standards passed with more than majority support, some standards did receive a number of important negative votes. Specifically, both DuMont and Philco expressed strong displeasure with the standards accepted by the FCC. DuMont was especially upset that the commission continued to reject the principle of flexibility, returning instead "with the same 30 frame, 441 line television which the Radio Manufacturers Association had accepted from its sponsor, RCA."[78]

The discussion in this chapter underscores how the development of technical standards for monochrome television in the United States was by no means a straightforward, rule-governed process based on the discovery of the one best system. When other countries set up television service, in a number of cases they authorized different standards. The FCC and the industry reached a final decision only after a prolonged period of negotiation and compromise, but these historical complexities tended to disappear behind the legitimating force of technical expertise.

As they had in the case of radio broadcasting, engineers during the crucial years of the 1930s and early 1940s played a major role in the formulation of public policy for television. The FCC demanded that a final decision should represent the best judgment of industry experts. It established the NTSC in response to this expectation that engineers ought to agree on a best set of standards. However, the standards debate raised a number of important questions and underscored the existence of important tensions. At least until it was forced to come to a final decision immediately before the war, the FCC, especially in comparison with the radio commission, took a more critical stance toward the role of technical experts in public policy and brought up a number of important issues about institutionalizing their expertise and objectivity. By examining the strategies used by the various participants in the standards debates and by placing them in the context of business strategy and New Deal activism, we gain a deeper understanding of the historical development of expertise and of decision making on policy for broadcasting during this period, especially the powerful role of boundary work and legitimation in providing closure to complex technical issues.

As a final point to this chapter, it should be pointed out that despite the critical attitude of Fly and the FCC toward the broadcast industry and earlier decision-making processes, the overall effect of the commission in the New Deal era was not radically to change the structure of the industry. In the case of television standards, despite Fly's efforts to keep RCA from dominating the new industry, the final standards chosen by the NTSC demonstrated the immense influence of this giant company. Further, the major attempt to restructure the radio industry—the chain monopoly investigation and the FCC's "Report on Chain Broadcasting"—did not lead to a radical reorientation of the network structure. The investigation had sought to evaluate the charge that "networks exercised a strangle-hold over affiliates, sucking up profits and preventing the local stations from exercising proper, localistic control over programming." Although the report found some truth in many of these charges and called for the restructuring of network/affiliate relations (the investigation forced NBC to sell off one of its two networks), the final rules issued by the commission, according to one authority, "had relatively little effect on radio broadcasting." The FCC tended to maintain conservative policies not necessarily because of "capture" by the industry but often because of congressional opposition to major change. Although an "independent" commission, the Federal Communications Commission cannot easily make major changes without support from the other branches of government.[79]

4)))

"Rainbow in the Sky"

FM Radio, Technical Superiority, and Regulatory Decision Making, 1936–1948

The Commission FM evangelists of yesterday, as today's leaders of the radio industry, seemingly have lost their zeal to bring to the people this utopia of broadcasting and listening potential. . . . FM channels in the sky go begging, and this new and superior radio service continues to be just a rainbow in the sky.

Commissioner Robert F. Jones of the FCC, January 17, 1950

When frequency-modulation (FM) radio was first developed, during the 1930s, its promoters—especially its inventor, Edwin H. Armstrong—were convinced that the new system's inherent technical superiority would guarantee its success in competition with the established amplitude-modulation (AM) system. W. R. G. Baker, an important leader of the radio-engineering community, argued that FM was "so much better technically than the present regular broadcast system that it can't fail of acceptance." Many radio engineers viewed the invention of FM as part of the "march of science which will obsolete the system now in use." They presented historical examples, including the triumph of AC electricity (alternating current) over DC (direct current), to drive home this point. In 1940, four years after the first public demonstration of his new FM invention, Armstrong confidently predicted that it would supplant the old AM system within five years.[1]

But nearly four decades passed before FM successfully challenged AM radio's supremacy in the United States. Not until after 1979 did FM's share of radio's listening audience exceed AM's. Historical studies that have examined the failure of FM broadcasting to live up to initial expectations generally repeat the story told by Armstrong and the FM pioneers, who argued that his invention was suppressed by the dominant commercial interests, especially the Radio Corporation of America (RCA) and its subsidiary the National Broadcasting Company (NBC). FM supporters charged that instead of working for the public interest, these companies were mainly committed to protecting their economic investment in the "inferior" system of network AM radio and in the development of the nascent television industry. Certainly the most serious charge was that the government agency responsible for regulating the broadcast industry, the Federal Communications Commission (FCC), was actively supporting big business's efforts to suppress FM. Armstrong's supporters portrayed him as "an individual warrior struggling against organized evil." The only recent book on Armstrong and FM is subtitled "One Man vs. Big Business and Bureaucracy." Armstrong's biographer contends that the "vast concentration of economic power" in the broadcast industry "rolled over FM and crushed it to a shape less threatening to the monopolistic pattern of operations."[2] Armstrong's suicide in 1954, at the end of the fifth year of a grueling litigation with RCA over the patent rights to the invention of FM, gives cogency and drama to this standard history.

There may be some truth in this view of Armstrong and the development of FM broadcasting, but the focus on a search for conspiracies tends to give a blinkered perspective that ignores complexities and fails to engage broader analytical and contextual themes. This chapter focuses on a crucial episode in the early history of FM radio. It involves one of the FCC's major decisions—a decision that Armstrong believed was motivated by the desire of both the AM radio industry and the television industry to severely cripple FM. As part of a new system of allocation for postwar utilization of the electromagnetic spectrum, in June 1945 the commission ordered FM stations to broadcast in the 88–106 MHz frequency spectrum, instead of in the lower 42–50 MHz band, where the industry had been operating since January 1941. This decision made the old FM system obsolete and forced the engineers, manufacturers, and broadcasters who had pioneered the industry to begin again from scratch and compete on an equal basis with RCA and other manufacturers who had not yet invested heavily in FM broadcasting. The commission's simultaneous decision to place one of the television channels in the old FM band only intensified the suspicions of the FM industry about the "hidden forces" at work "behind the commission's actions." Armstrong bitterly denounced the new allocation, which the FCC justified on purely technical grounds, as "one of the major mistakes in engineering history." He also argued that "in their attempt to preserve and extend the monopoly of broadcasting, the chains [AM networks] . . . enlisted the support from some of the strongest political forces in the country."[3]

A detailed examination of this decision and an exploration of its larger implications is especially important for illuminating the intersection of broadcast technologies and public policy during this period—specifically, the interplay between technical problem solving and economic, social, and political decision making. After briefly discussing the invention of FM radio and the policy decisions of the FCC during the 1930s and early 1940s, this chapter analyzes the 1945 allocation decision, focusing on the complex negotiations among different institutions and individuals whose involvement helped shape the new technology. I examine key aspects of the decision-making process, including the attempts by the FCC to legitimate its actions to the public and the response of opponents to the new allocation.

A major theme that provides a framework for understanding the particular strategies used by both opponents and proponents of the decision is the tendency in regulatory policy making to apply technocratic values. Although a few participants in the debate over the 1945 allocation emphasized that technical evaluation invariably involved social, economic, and political considerations, many participants held strongly technocratic views. They generally wanted to delegate primary responsibilities to engineers and make a clear distinction between policy decisions and technical evaluation. But the resulting controversy involving conflicting engineering evaluation underscored fundamental problems in this effort to draw sharp boundaries, as well as in the notion that innovations possessed intrinsic technical superiority. An analysis of the strategies used by participants in the public policy debates thus helps clarify the complex forces shaping the early history of FM radio and underscores how closely support for FM was connected to the early enthusiasm for television.

The Early History of FM

Armstrong received a patent for wide-band frequency modulation in 1933. His earlier inventions (including the regenerative "feedback" circuit, the super-heterodyne circuit, and the super-regenerative circuit) had played a crucial role in fostering the radio broadcast boom of the 1920s. Armstrong's chief motivation for developing FM was to eliminate the problem of natural and man-made static, which had plagued radio since its early development in the late nineteenth century. Because AM radio waves and the electrical signals that produce static have similar propagation properties, AM radio receivers are unable to discriminate between the two kinds of signals.

The modulation of broadcast signals refers to the way information is superimposed on a carrier wave of a particular frequency. AM broadcasting adds messages by varying the strength or amplitude of the wave. FM encodes information by changing the wave frequency. Armstrong began investigating frequency modulation in 1925. Earlier experimenters believed that they could reduce static only by narrowing the broadcast channel. This had been appropriate for AM, but when researchers treated FM in this way they found it unsuitable for radio broadcasting. In 1932, Armstrong discovered that he could greatly reduce static by widening the band of frequencies used. The key patents Armstrong received in 1933 covered the development of transmitters and receivers for his wide-band FM system.[4]

Beginning late in 1933, Armstrong received support from RCA to improve his system. Armstrong's close connections with RCA dated from 1922, when he sold the patent rights on his super-regenerative circuit to the company, became RCA's largest individual shareholder, and promised RCA first option on any new invention. The cordial relationship ended in the spring of 1935, when officials asked Armstrong to remove his FM equipment from the NBC station in the Empire State Building so that the company could concentrate on its television experimentation. RCA's decision to invest in the development of television rather than FM left Armstrong suspicious about any decision made by the broadcast industry that did not seem to support FM. Armstrong actually acknowledged that RCA had a right to make this kind of business decision. He was mainly upset with what he believed were illegitimate business practices, including what he called a "talk down campaign" against FM, improper lobbying of government regulators, and a misrepresentation of engineering facts.[5]

Armstrong gave the first public demonstration of FM broadcasting in 1935, soon after his break with RCA, at a meeting of the Institute of Radio Engineers. Although the members of the institute seemed impressed by this demonstration, Armstrong was disappointed that this initial enthusiasm did not lead to overwhelming public and private support. Indeed, within a year after this event, Armstrong believed he saw forces working against his invention. The FCC's 1935 annual report (issued in January 1936) contended that stations broadcasting in the higher frequencies where FM would have room to operate (above 30 MHz) "would serve only a few miles, probably in the order of two to ten miles." The report made no mention of Armstrong's tests that demonstrated transmissions over distances of more than eighty miles. Armstrong was especially suspicious when, a few weeks after writing this report, Charles Jolliffe, the FCC's chief engineer, accepted a position in charge of RCA's frequency bureau. Armstrong insisted that Jolliffe, despite his denials, must have been aware of these tests.[6]

Jolliffe's main responsibility at RCA was to obtain the best frequency allocation for television. In the spring of 1936, the FCC requested technical information from the industry to help it decide how to develop the newly available higher frequencies above 30 MHz, where both FM and television would operate. Jolliffe's representations at this meeting further convinced Armstrong that RCA was actively working to suppress FM. Instead of informing the commission about the extensive FM tests RCA had sponsored, Jolliffe made no mention of FM, although he did discuss in general terms "high frequency broadcasting," which he later claimed included FM.[7]

Armstrong charged that RCA used the "promise of television" to "create a shortage of channels" for FM. He believed that RCA's decisions involving the development of television were largely motivated by a desire to protect its investment in AM network radio (including its control of NBC and key AM radio patents) from the threat of FM competition. In comparison with television, which was years away from extensive commercial use, according to Armstrong, FM was an established technology and was more deserving of support. In May 1936, the FCC assigned FM exclusive use of an approximately 1 MHz band in the vicinity of 42 MHz, or enough spectrum space for four channels. It gave experimental television exclusive use of more than 50 MHz, or enough room for eight channels. Armstrong complained that "the promoters of television" received a "virtual monopoly of the frequency bands."[8]

By the late 1930s, Armstrong believed he had conclusively demonstrated the technical superiority of FM. Most significant was FM's ability to eliminate naturally produced static and most man-made interference. Further, FM stations broadcasting on the same frequency were much less likely to interfere with one another than AM stations under the same conditions. With Armstrong's FM system, interference did not occur until the interfering signal was half as strong as the signal from the desired station. An AM signal created interference even when it was one-twentieth the strength of the desired signal. Finally, FM could transmit high-fidelity sound reproduction.[9]

Armstrong's troubles became potentially more serious late in 1939, when RCA requested the FCC to make television's temporary experimental channels permanent. If the commission had granted this request, FM would have been "boxed into" an inadequate band of four channels, without any unused frequencies available in adjacent bands for future growth. By this date, more than one hundred stations had already applied to broadcast in this limited band. Armstrong believed FM needed additional spectrum space not only to accommodate these new applications but also to stimulate interest among other potential investors. Two of RCA's main competitors took an early interest in FM. General Electric Company (GE) and Zenith Radio dominated the early FM-equipment market. Both companies also began operating FM stations in 1939, a few months after Armstrong's Alpine, New Jersey, station—the first "full-powered FM station"—went on the air. Another Armstrong supporter, the Yankee Network of New England, made the first attempt to set up a network of FM stations in 1939. In 1940, Armstrong and the other FM broadcasters, including a number of newspaper publishers, came together to form an FM trade association, FM Broadcasters, Inc.[10]

Hearings by the FCC in 1940 favored FM, thanks mainly to the support of the new chairman, James Lawrence Fly. As we have seen, Fly believed the FCC should not limit itself to evaluating technical issues, but argued for an "integrated and comprehensive regulatory policy" that took into account important social, economic, and political concerns. Fly saw the growth of FM radio as a way to limit monopoly control by dominant elements in the radio industry, especially RCA and NBC. Armstrong's disclosure of confidential RCA engineering reports on FM experimentation that the company did not present to the commission in 1936 helped convince Fly to decide against RCA's allocation request. By transferring television's number one channel to FM (44 to 50 MHz), the final allocation gave FM a total of forty channels in which to expand. The FCC also authorized commercial development, which went into effect in January 1941. When the United States entered World War II in December, the FCC had authorized sixty-seven commercial FM stations, and forty-three applications were pending. Contemporary sources disagreed on the number of receivers in public use at the beginning of the war, but the generally accepted figure was five hundred thousand. Thus, despite FM's early problems, the industry was, according to Armstrong, "going great guns at the time of Pearl Harbor." Although the federal government placed a freeze on the civilian electronics industry at the beginning of the war, officials allowed FM stations already broadcasting to continue operations.[11]

Armstrong's charges of conspiracy and dishonest practices by big business and government regulators form an important part of the history of FM; the purpose of this chapter is not, however, to evaluate definitively the accusations from this early period. All of Armstrong's charges cannot be proved conclusively by a study of surviving archival material. Jolliffe's personal papers and relevant RCA records have apparently not been preserved. Although some convincing evidence to support his position exists, Armstrong, for his part, tended to present conflicts and disputes in less than subtle, black and white terms. More important for this study is recognizing how these early debates provided a framework for later developments, especially by predisposing Armstrong to suspect individuals and institutions of working against his new invention.

FM and the Radio Technical Planning Board

To understand the origin of the commission's 1945 decision to shift FM's frequency allocation, we need to look at plans begun during the war to establish standards for the postwar civilian electronics industry. The industry pressured the FCC to make a decision in a timely manner to give the industry sufficient time to prepare for postwar expansion and the State Department enough time to prepare for postwar international agreements on the use of the electromagnetic spectrum. Wartime research stimulated new developments in electronics, including new tubes and circuits, which helped open up higher frequencies to commercial exploitation. As early as 1942, industry and government officials recognized that extensive planning would be needed to develop a new allocation scheme for the use of frequencies above 30 MHz (AM radio operated in the 550 to 1600 kHz band). In November, during a joint meeting of the Institute of Radio Engineers and the engineering department of the Radio Manufacturers Association, FCC chairman Fly encouraged the radio industry to establish an organization that would work to hasten reconversion to peacetime production and employment by providing the commission with the necessary engineering advice for developing frequency allocations and system standards. He suggested setting up a new group modeled on the National Television System Committee (discussed in chapter 3). Fly's proposal resulted in the establishment, in September 1943, of the Radio Technical Planning Board (RTPB). At least eighteen "nonprofit associations and societies" sponsored the board, including not only professional engineering and trade associations like the Institute of Radio Engineers and the Radio Manufacturers Association but also broadcast groups like FM Broadcasters, Inc., the National Association of Broadcasters, and the Television Broadcasters Association.[12]

The responsibilities of the planning board were much more extensive than those of the television system committee. During 1944, six hundred board members conducted work divided among thirteen panels. Panel two sought to coordinate the use of the entire frequency spectrum and reconcile conflicting frequency allocations recommended by different panels. Panel five was responsible for specific recommendations for FM broadcasting, including both frequency allocations and system standards. Other panels studied and developed standards for such services as television, facsimile, standard AM broadcasting, and aeronautical radio. The board sought to include the most competent "specialists in radio propagation" as well as "any individual or organization having either a direct or indirect interest in any of the services or problems to be considered" by the planning board. Significantly, the engineers with the Institute of Radio Engineers and the Radio Manufacturers Association who established the board "restricted" the analyses and recommendations of the panels "to engineering considerations."[13]

The FCC pressured the board to provide recommendations as soon as possible. The commission was not only concerned about making sure the civilian electronics industry got off to a quick start when the war ended, but also needed to provide the State Department with a comprehensive frequency-allocation proposal so its telecommunications division could be ready for the International Telecommunications Conference to be held immediately after the war. The FCC was mainly responsible for allocating domestic frequencies used by nongovernmental services. The Interdepartment Radio Advisory Committee, which included representatives from different government agencies using radio, coordinated government use of radio.[14]

Panel five of the board met between December 1943 and June 1944. The chairman, Cyril Jansky, had overseen the construction of the first FM station in Washington, D.C., and had served as president of the Institute of Radio Engineers. The technical problem that would ultimately play a major role in the FCC's decision to shift FM's frequency allocation, the potential for sky-wave interference, became a major topic of discussion for members of panel five, especially during its second meeting in April 1944. As early as the preceding April, a commissioner had expressed concern about the problem of sky-wave interference in the vicinity of 40 MHz, the band in which FM operated. Actually, officials had raised doubts about this location for FM in 1940, when the FCC first authorized commercial operation. Three years earlier, Armstrong himself had admitted that "the indications are that there will be much less trouble at 100 megacycles than on 40 megacycles."[15]

FM broadcasters worried about the potential for two kinds of sky-wave interference: F2-layer transmission and sporadic E transmission. F2-layer interference occurred when transmitted waves from distant stations (often more than one thousand miles away) reflected off the upper (F2) layer of the ionosphere and interfered with stations broadcasting on the same frequency. Engineers agreed that this type of interference decreased with increasing frequency. Under normal conditions, signals above 40 MHz were not reflected by the F2 layer of the ionosphere; for these higher frequencies, propagation occurred by direct line-of-sight transmission. During the hearings conducted by the RTPB and the FCC in 1944 and 1945, some engineers warned that F2-layer reflections might occur at higher-than-normal frequencies during maximum sunspot activity. Experts believed that sporadic E interference occurred when signals were reflected by irregularly distributed areas of ionization in the intermediate (E) layer of the ionosphere. Engineers and broadcasters thought this problem was more prevalent during the summer months.

Interference was also known to occur in the lower regions of the atmosphere. Engineers worried about various tropospheric effects, including long-distance bending and ducting of waves. They thought this form of interference increased with higher frequencies. Signal shadows were also known to occur behind hills and buildings. This form of interference similarly seemed to become more noticeable at higher frequencies.[16]

During the April meeting of panel five, William Lodge, director of engineering at the Columbia Broadcasting System (CBS), announced that, despite having voted during the first meeting in favor of keeping FM in its current band, he now believed evidence indicated that during the next sunspot maximum, sky-wave transmission might create intolerable interference in the 40 to 50 MHz band. Three other members also expressed reservations about these lower frequencies and told of specific cases when observers had reported sky-wave interference. NBC engineer Raymond F. Guy reported that receivers in the United States had picked up European television stations broadcasting on the same frequencies. All members agreed, however, that they lacked the necessary data to make a definitive decision. Very few stations had existed during the last sunspot maximum, and engineers had made even fewer observations under normal conditions at higher frequencies around 100 MHz. The chairman emphasized the inherent technical uncertainty of the decision, insisting that if they wanted to wait until all the facts were in they would "never make a decision."[17]

Following a suggestion made by Lodge, the panel decided to defer this interference problem to John Howard Dellinger, chief engineer of the Bureau of Standards and probably the most important engineer in government service. They considered Dellinger one of the foremost authorities on radio propagation and believed his group at the bureau would have the most extensive and reliable set of data. Dellinger responded in general terms that the fear of long-distance, sky-wave interference in the 40 to 50 MHz band "is not well founded." Although he believed no good reason existed to shift FM to higher frequencies, he also emphasized that "no frequencies are free from transmission vagaries." The panel voted seventeen to three to keep FM in the lower band after receiving Dellinger's letter. Lodge voted in favor of the recommendation. Two of the engineers who voted against the proposal felt officials needed to collect more data before they made a final decision; the third engineer, Thomas Goldsmith of DuMont Television, remained convinced that his data justified moving FM to a higher band. Panel five's final report, dated June 1944, recommended that FM stations continue to broadcast in the vicinity of 40 MHz, but in an expanded band; there would be not forty, but eighty to one hundred channels, each 200 kHz wide.[18]

Commission Hearings, Fall 1944

The FCC held official hearings beginning on September 28, 1944, in order to allow public presentation of all available evidence, including the recommendations of the planning board, concerning the allocation of the entire frequency spectrum, from 10 to 30,000,000 kHz. The trade publication *Broadcasting* characterized the commission's plans for postwar allocations as the "most sweeping revision of the radio spectrum since the art began." More than two hundred witnesses testified at the commission's hearings, which lasted twenty-five days. Government officials recorded 4,559 pages of testimony and received 543 exhibits from industry engineers, government engineers, business leaders, and other individuals interested in telecommunications policy. The FCC felt pressured to develop an allocation plan in a timely fashion in order to meet a December 1 deadline established by the State Department. Contemporary observers argued that a "race against time" was apparent in the proceedings.[19]

The FCC first issued orders for the September hearings four days after the State Department held its own conference, on August 11 and 12, to help plan for international allocations. Dellinger, the chairman of the State Department's technical subcommittee on telecommunications presided over the conference. The Interdepartment Radio Advisory Committee (of which Dellinger was a member) presented a preliminary allocation that reflected the needs of the federal government. The interdepartmental committee recommended sixty channels for FM in the 42–54 MHz band, but also indicated that technical studies in progress might eventually justify a shift to higher frequencies. In formulating its own plans, the FCC also needed to consider this proposal, but the State Department emphasized that there was room for flexibility.[20]

Before the start of the commission's hearings, two conflicting proposals for the FM allocation were reconciled through behind-the-scenes negotiations. The proposal developed by panel five of the RTPB partly conflicted with the recommendation panel six presented for the placement of television's first channel. The chairman of panel two, Charles Jolliffe, met with members from each panel and with the pro-FM chairman of the planning board (the above-mentioned W. R. G. Baker, who headed the electronics division at GE) in order to mediate a major dispute between FM and television. After making "full use" of the proposal of the interdepartmental committee and the advice of a government representative, the planning board made a final recommendation during the commission's hearings in September that FM receive seventy-five channels in the 41–56 MHz band.[21]

Despite this recommendation, which was supposed to represent the best advice of the radio-engineering community, it became clear during the hearings in early October that the FCC might rule against the planning board's technical experts. The commissioners and the chief engineer began questioning witnesses about moving FM to frequencies in the vicinity of 100 MHz. The main source of support for the move at the commission hearings came from Oliver Lodge of CBS and T. A. M. Craven,

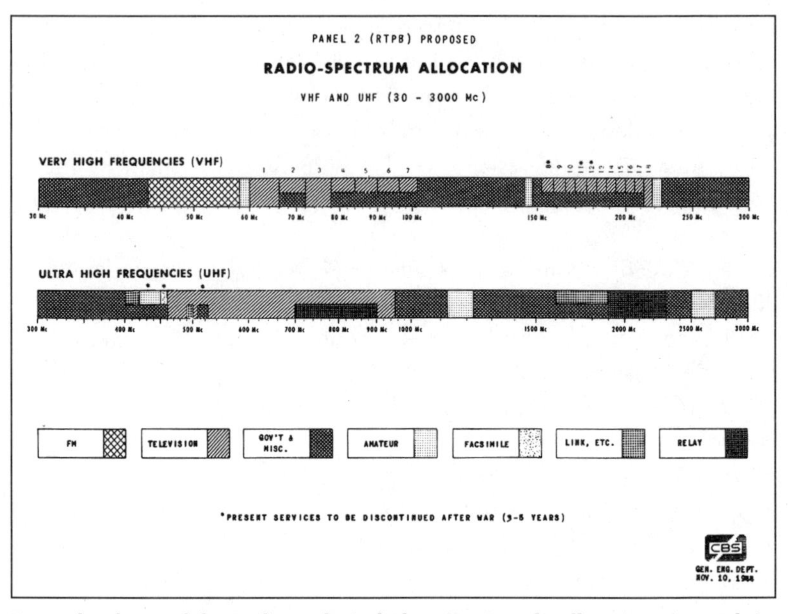

Example of one of the Radio Technical Planning Board's allocation proposals in 1944 for VHF and UHF.

a former commissioner (and one of the few commissioners trained as an engineer), who was now representing Cowles Broadcasting Company. On August 14, about two months after voting in favor of the recommendation of panel five to keep FM in its lower band, Lodge had published an article in which he again warned of the dangers of sky-wave interference. Without presenting many details, he claimed that new tests in July demonstrated the existence of serious E-layer interference; Lodge also repeated the old warning of F2-layer interference during the next sunspot maximum—this time without including any new data.[22]

FM industry representatives responded angrily to Lodge's article and his support of the frequency shift. Armstrong later pointed to Lodge as the main source for the idea of moving FM. FM supporters accused the CBS engineer of using the technical issue of interference as a "smoke screen" to maintain the dominance of AM network radio. But Lodge's motivations should not simply be analyzed in the context of competition between FM and AM, or even between FM and television; his desire to shift FM upward had more to do with internal conflicts within the television industry.[23]

Both CBS and Cowles Broadcasting wanted to move television broadcasting to much higher frequencies (above 300 MHz instead of in the vicinity of 100 MHz) where the new high-definition, color system the companies were trying to develop would have room for growth. Most of the remainder of the television industry, especially RCA/NBC and DuMont Television, lobbied the commission to protect their investment in the old television system in the lower frequencies. The proposal from CBS and Cowles to move FM into frequencies being used by television should be seen in terms of their desire to disrupt the established television system and the economic interests supporting that system. As far as CBS and Cowles were concerned, the shift was for FM's own good, since the plan would give the industry room to grow—from a congested band to a wide-open region where it could, presumably, compete against AM stations (once television also had been moved upward). Testifying for Cowles Broadcasting, Craven proposed that FM should have as many as 400 channels in the vicinity of 100 MHz.[24]

The commissioners were receptive to the proposals of CBS and Cowles because they were convinced of the advantage of a policy that would result in expanded frequency bands for both television and FM. Although Fly left the commission in November 1944 before it had made a final decision, his line of questioning during the hearings indicated he supported shifting both FM and television to higher frequencies, where both services would have room to grow.[25] Fly had been the strongest supporter of FM in 1940, and it seems unlikely that he would have favored a policy that might hurt the market position of the new technology. On the contrary, he initially believed the FM shift would help strengthen the industry.

The idea for shifting FM upward gained momentum during the autumn of 1944 because of the generally low-key response from Armstrong and other FM supporters at the commission's hearings. They did not seem to think an all-out lobbying effort was necessary to defeat a technical policy decision that they believed lacked the support of the engineering community. Further, the particular line of questioning pursued by the commission seemed to force representatives of the FM manufacturers, notably Zenith and Stromberg-Carlson, to acknowledge that sky-wave interference would be less important in higher frequencies and that the industry would probably be able to survive a frequency shift. Armstrong, for his part, seemed to have been preoccupied fighting a proposal to reduce the channel width of FM broadcasts from 200 to 100 MHz.[26]

Engineering testimony presented toward the end of the commission's hearings in 1944 gave the FCC what it believed was "authoritative" technical evidence to support a decision to shift FM. By far the most important technical evidence, and the most controversial, came from Kenneth Norton, an engineer with the FCC who had also served during the war with the operations and analysis division of the War Department. Norton presented detailed graphs and charts of propagation data from the National Bureau of Standards laboratories near Washington, D.C., and measuring stations in other parts of the world. The U.S. Army-Navy Interservice Radio Propagation Laboratory collected most of the worldwide data (at this time, data was subject to wartime restrictions). Norton argued that the available data "demonstrated the necessity for moving FM upwards in the spectrum" because it indicated serious problems from both sporadic E-layer and F2-layer interference for frequencies as high as 80 MHz.[27]

Despite expert testimony by engineers on the RTPB against this recommendation, the FCC's proposed allocation, announced in January 1945 after further consultations with the government's interdepartmental advisory committee for radio, placed FM in the 84 to 102 MHz band. The proposal retained the 200 kHz band width, giving FM ninety channels in which to operate (instead of forty). The commission also included provisions for FM's expansion upward into 102–108 MHz and downward into 78–84 MHz. Thus, FM might potentially end up with 150 channels. Because of the demands of competing services, especially civil and government aviation, the proposal reduced the number of television channels from eighteen to twelve, placing them in two bands of the remaining frequencies between 44 and 210 MHz. Television's lower band would include FM's old frequencies, but the FCC specifically wanted the industry to consider the two television bands as temporary allocations that it would use only until it was ready to move television into the ultrahigh frequencies (UHF) above 480 MHz. The FCC also announced that the proposed allocation would not go into effect until groups and individuals with an interest in the decision had an opportunity to express their views in a series of public hearings.[28]

The television allocation was a compromise between the CBS and NBC/RCA proposals. Since CBS won a promise from the commission to expedite a shift to the UHF spectrum, the decision to continue television broadcasting in the old frequencies was only a partial victory for NBC and RCA. But RCA's success in continuing low-band television, at least temporarily, also seemed related to the proposal to move FM. The manufacturers who favored immediate authorization of television in the prewar channels had testified, especially during the allocation hearings held by the government interdepartmental committee, that they preferred "to get as many channels as we can down low and get started." They pointed out that tubes and other equipment necessary for transmitting were available only for the seven lowest prewar channels, all below 108 MHz. Manufacturers and broadcasters also preferred the lowest channels, as near to 40 MHz as possible, because they believed shadows and multipath interference would become more of a problem in the higher frequencies. Jolliffe, RCA's chief engineer, testified that "if the number below 100 MHz could be increased, it would be better." Although they did not argue the fact publicly, the television representatives allied with RCA must have recognized that moving FM out of this lower band would help maximize the number of channels available for television. They believed television, rather than FM, would be the key industry stimulating consumer spending and creating jobs in the postwar period: "FM is merely an improvement in an existing system. . . . It is going to take a new service—a different service—something which they don't now have to bring out a great deal of enthusiasm [among consumers]."[29]

Of all the different industry groups, FM supporters were the least satisfied with the January 1945 allocation proposal. But it is important to recognize that Fly and other members of the commission initially favored the shift as a way to help the FM industry. As soon as the FCC followed through on its promise to shift television "upstairs," FM would have room to expand and grow. Specific technical testimony thus helped legitimate a decision that had originally been formulated to help the FM industry.

Technocratic versus Nontechnocratic Decision Making

The exclusive use of technical criteria to legitimate complex policy decisions underscores the tension in the work of regulatory commissions emphasized in previous chapters. The wish both to delegate authority to technical experts to avoid conflict and to stress broad democratic judgment to take into account the controversial socioeconomic factors of standards decision making has been a central dilemma in technology policy making. This tension is especially evident in the way the commission dealt with the FM allocation proposal of January 1945.

In arriving at this proposal, the FCC considered the economic impact of the frequency shift and concluded that the cost to manufacturers, station owners, and the public would "not be great." The official report on the proposal acknowledged that the "determination was not limited to technical considerations but also took into account economic and social factors and considerations of national policy." However, when commissioners and staff members defended this decision in controversial public forums, for instance before the House Appropriations Committee, they emphasized that the FM shift was demanded purely by technical considerations. The secretary of the commission responded to public inquiries by bluntly arguing that "the reason behind the Commission's proposal to move FM higher in the spectrum is that engineering data, some of it available for the first time, shows that FM would be subject to intolerable sky-wave interference if it remained at its present assignment and that no such interference would be expected in the higher portion of the spectrum."[30]

In order to understand why the FCC used exclusive technical arguments to legitimate a decision that had also taken into account nontechnical considerations, we need to consider the broader political climate of the period. During the three years before the announcement of the allocation proposal, the commission had been the focus of intense controversy, mainly because of the activist policies of chairman Fly. The journal *Broadcasting,* which tended to reflect the views of the dominant elements in the broadcast industry, complained in 1944 that "probably no Government official in our times, has used more intemperate or abusive language in dealing with industry."[31]

Fly's actions antagonized not only the radio networks but also conservative members of Congress, who, beginning in 1941, launched a series of investigations of the FCC. Their charges ranged from general complaints that the commission was "acting arbitrarily and exceeding its powers" to specific attacks on employees, who were characterized as "un-American" subversives. Members of Congress introduced a number of legislative bills—albeit unsuccessfully—to revise the 1934 Communications Act. Technocratic advocates of "free radio," such as the Republican presidential candidate Thomas E. Dewey and the influential consulting engineer John V. L. Hogan, called for "unambiguous" legislation that would restrict the FCC to regulating the technical aspects of broadcasting, instead of "debatable" concerns such as programming, business and economic policies, and station and network relations. An important "debatable" concern was the commission's decision, during the late 1930s,

Drawn for BROADCASTING by Sid Hix

"We have a little problem here, Dr. Einstein, and we were wondering . . ."

Cartoon representing the complexities of radio frequency allocation. It also demonstrates how officials were inclined to follow technocratic values and try to delegate responsibilities to technical experts. From *Broadcasting* (March 18, 1946): 16.

to reserve a number of channels in the FM band for educational broadcasting. Educators had first asked for this special consideration during the late 1920s, when the Federal Radio Commission began regulating AM radio. But the exclusive technical criteria used by the FRC had assumed that nonprofit, educational stations were no more valuable than commercial stations. Fly's rejection of this technocratic position by working to set aside a band of FM frequencies for noncommercial broadcasters brought further criticism from his opponents. *Broadcasting* complained that "once again the commission ventures into social and economic stratospheres which are questionable."[32]

The two chairmen who succeeded Fly and who were responsible for implementing and defending the 1945 allocation had closer ties to the radio industry and were less willing to pursue activist policies. The attorney who replaced Fly late in 1944, Paul Porter, had been employed for a number of years by CBS. Charles Denny, who succeeded Porter in 1946, also after working for CBS, resigned after one year to become NBC vice president and general counsel. The strategy of technical legitimation pursued by the commission to justify the 1945 allocation reflected a new policy of avoiding controversies by retreating from earlier nontechnocratic, activist practices.

The tension between technocratic and nontechnocratic policies had also been an important factor in decisions made by the RTPB. Although the board's bylaws restricted the activities of the different panels to "technical" considerations, some members criticized this position as unrealistic and overidealized. When questioned whether the board would take into account economic investments that organizations might have in different parts of the spectrum, chairman Baker, responded in surprisingly candid terms: "I consider that part of the engineering problem. I don't differentiate between the economic and purely technical." Members of the board's panel five specifically argued that "a question of allocation must, to some extent, give consideration to . . . other policy matters." The chairman of panel five, Cyril Jansky, believed that this was especially true for decisions about the number of channels assigned to different services, which he characterized as "not purely an engineering matter but one which in fact is primarily . . . a question of public policy."[33]

An analysis of the record indicates that in formulating decisions, panel five did take into account social and economic factors. When panel members rejected moving FM to a higher band of frequencies, they not only based their decision on the technical evidence but also on "the fact that there is already a substantial public investment in FM equipment and a highly organized public service already being rendered by existing FM stations in this position of the spectrum." Complex, hybrid decision making was necessary because of the technical flexibility and uncertainty of the work. Jansky emphasized that "no panel or no group will know all there is to know about all of the frequencies in the band which we are studying." At least one engineer on the planning board argued further that the organization should include individuals who were not strictly professional engineers because "the factors before the Committee are not all engineering factors."[34] Like the FCC, the board thus legitimated decisions that involved complex considerations through exclusive reference to technical criteria. In the case of the board, because of the restrictions imposed by the founders, this tension was inherent in the structure of the organization.

After announcing the proposed allocation in January, the commission gave individuals and institutions an opportunity to respond during hearings from February 28 through March 1. Opponents of the FM shift mobilized to fight the decision during the month preceding the hearings. The major parties with an interest in the allocation—including engineers, station owners, network executives, manufacturers, and trade associations—responded to the decision in briefs submitted to the commission. Thirty of these representatives also testified as witnesses at the February-March hearings. A "secret hearing" was then held on March 12 and 13, during which the participants discussed the classified military data used to help justify the commission's allocation proposal. Dellinger appeared before the commission at this proceeding and reiterated the position he had taken in May 1944 against the FM shift.[35]

After the March hearings, the FCC delayed making a final decision until June 27. The State Department had extended the original December deadline it had given the commission, and by the spring of 1945 it was satisfied with some of the allocation decisions that the commission had already made for other parts of the frequency spectrum. The delay in the decision on the allocation of FM and television also occurred because of an announcement in May by the War Production Board that the freeze on the civilian electronics industry would continue until military cutbacks reached 75 percent, which was not expected until at least the first quarter of 1946. The board assured the commission that it would give a ninety-day notice before lifting controls. A major consideration for the FCC had been to develop a new allocation quickly in order to give the industry enough time to prepare for postwar development. Because of the announcement by the War Production Board, the FCC now believed it had sufficient time to conduct further engineering measurements. Although the engineers on the commission thought that they already had enough technical evidence to justify moving FM, at least two of the commissioners were convinced that engineers needed to make more observations. The May 14 issue of *Broadcasting* reported that the commission was "in a three-way split over FM." An announcement in May reflected this disagreement; new engineering tests would help the commission decide between three alternative allocations for FM: (1) 50–68 MHz; (2) 68–86 MHz; and (3) 84–102 MHz.[36]

Early in June, the War Production Board reversed its earlier decision and announced that the government would lift the freeze as soon as the Japanese surrendered; they also warned not to expect a ninety-day advance notice. This statement shocked the radio and television industry into pressuring the FCC to make a decision immediately on the allocation of radio and television. The Radio Manufacturers Association, the Television Broadcasters Association, and FM Broadcasters, Inc., all warned that further delay might result in postwar unemployment since manufacturers needed a significant period of time to design and produce new transmitters and receivers for operation in the new frequencies. The three major trade associations also urged the commission to adopt the 50–68 MHz allocation for FM. The FM supporters felt confident about pressuring the FCC at this time because they mistakenly believed commissioners were ready to choose a lower band. On June 12, the president of Zenith Radio wrote a friend that "it looks as though [the FCC is] going to compromise on the 50 to 68 megacycle band which *is acceptable* and will not cripple FM." After the commission held a hearing on June 22 and 23, it took only three days to make a final decision to reject the recommendation of the FM industry and not wait for further engineering measurements but proceed with the original proposal and move FM to the 88–106 MHz band. In August, when the commission moved facsimile service from a temporary allocation in the 106–108 MHz channel, FM achieved its full range of frequencies from 88 to 108 MHz. The long and hostile dispute had helped polarize engineers from the FCC and the industry planning board. The final decision reflected the influence of the FCC engineers, who had too much at stake, including pride and professional standing, to agree to a compromise.[37]

There is also evidence that key FCC engineers felt television was more important than FM and deserved preferential treatment. A memorandum dated February 1945 listed a number of reasons "why television needs the lower frequency channels": receivers and transmitters were available for use only in the lower frequencies, the lower frequencies were less vulnerable to shadows and multipath interference, "better coverage of service areas" would occur with the lower frequencies, and there would be "no image interference and less drift problem if FM is on high side of Main Television channels." The memorandum also contended that "the growth of this important industry should be favored, not compromised" and "the public should be given television without further delay." Although a handwritten note on the document stated that "this represents the thoughts of most of the Television and FM engineers" in the FCC's receiver division, the unidentified official also wrote that "however for policy reasons it cannot be accepted as official." The potential threat of sky-wave interference thus seemed to provide a clear technical rationale for legitimating a decision that was based on a number of complex concerns.[38]

Critics of the FM shift wanted to know why the commission did not seem to take into account the fact that television would also be affected by sky-wave interference that might occur in the old FM band. FCC engineers countered that television would have fewer stations "at the proper geographical distance to bring in the long-distance interference." And since FM, unlike television, "would be programmed full time at the outset," FM broadcasts would have more chances of being affected by interference. They also considered FM as more vulnerable than television because it was a permanent service. Officials assumed that television would probably be moved again to a higher band well above 300 MHz. Finally, Norton argued that FM would actually benefit from the move to 100 MHz, because the higher frequencies would allow the stations to serve larger areas, especially rural regions, with interference-free signals.[39]

The historical contingencies that helped shape the allocation debate give us only a partial understanding of how officials made the final decision. We also need to take into account the tension between technocratic and nontechnocratic views, which played a central role in the decision-making process. Although the record clearly indicates that the commission took into account "nontechnical" factors, the FCC continued to justify publicly the FM shift by referring to the technical criteria of engineering testimony. The technocratic legitimation strategy pursued by the FCC after January 1945 took three forms. The first continued to emphasize the scientific evidence that guided the commission's decision. Chairman Porter argued in March 1945, for example, that the "rightness of this decision turns upon an evaluation of engineering data," which demonstrated the existence of interference in the lower frequencies. The second form of legitimation emphasized the scientific, disinterested authority of engineers who made the decision. Thus, Porter also insisted that the commission was guided "by the recommendation of our technical staff whom I believe to be competent, disinterested, and without any private axe to grind." The unstated assumption behind this statement was that the engineers on the commission were more reliable because, unlike the RTPB engineers, they did not have a vested interest in any aspect of the broadcast industry. Indeed, the engineers on panel five who opposed the FM frequency shift mainly represented FM broadcasters and manufacturers. The final method of legitimation was used in 1948 by the newly appointed chief engineer of the FCC, George Sterling, who was responsible for justifying the decision to Congress, despite not having been involved in its formulation. Rather than primarily emphasize the technical evidence that pointed to the existence of interference, Sterling stressed the legitimacy of the process that the FCC had used to evaluate technical criteria and expertise. In testimony before the House Committee on Interstate and Foreign Commerce, Sterling emphasized that he mainly wanted to give "a clear and complete picture of the path the Commission followed in arriving at its decisions; of the opportunity that all parties had to appear; present testimony, and engage in oral argument; and of the full extent to which the significant factors involved in that decision were considered by the Commission."[40]

Despite the public statements of the commission, however, published and unpublished sources indicate that, during 1945, the evaluation process continued to include both "technical" and "nontechnical" considerations. At times, commissioners acknowledged that because "many of the factors involved a judgment upon abstruse technical considerations concerning which there is but little factual information," they needed to take into account nontechnical criteria. Most important, the commission tried to predict the economic cost of the FM frequency shift by asking manufacturers to evaluate the relative expense of producing equipment for the higher band as opposed to the lower band. Commissioners also inquired into the amount of time it would take manufacturers to convert to a new production system. The commission wanted to judge the contention of the FM manufacturers that the shift would cost the public millions of dollars in obsolete receivers and contribute to unemployment by delaying the resumption of the civilian electronics industry during a crucial period when returning soldiers would be looking for jobs. Expectations were high that FM radio would take off after the war, replace the AM system, and contribute to a postwar boom in the electronics industry. Even CBS, which had spearheaded the effort to shift FM upward, predicted that FM would soon "supplant" AM radio. Although the FM manufacturers testified that it would take up to two years or more to convert FM receivers and transmitters, the commission based its final decision on the testimony of Philco and other companies that it would take no longer than four months. The commission also emphasized the availability of converters that would allow the old FM sets to receive broadcasts in the higher frequencies, thereby preventing complete obsolescence. Further, the commission sought to mitigate the negative effects of the shift by establishing an interim period during which broadcasters could continue using the lower frequencies until they were ready to convert.[41]

The tendency of decision makers to attempt to follow technocratic views also provides an important framework for understanding the response of individuals and institutions, especially manufacturers and broadcasters, to the proposed FM shift. The supporters of the new allocation—including the three networks CBS, ABC, and Cowles Broadcasting—emphasized that their evaluations were based on technical, rather than economic or social considerations. Manufacturers who testified or presented written briefs favoring the move included companies that, in general, had not yet invested heavily in FM: DuMont Laboratories, Majestic Radio and Television Corporation, Hallicrafters Company, Philco Company, and Crosley Corporation. In most cases, engineers presented the positions of their companies. Two noncorporate groups also supported the shift—amateur radio operators and the International Association of Police Chiefs. Both groups believed the new allocation better served their interests. Under the new allocation, for example, the commission allowed the amateurs to use a band 4 MHz wide as opposed to the 2 MHz recommended by the RTPB.[42]

Opponents of the FCC Policy

Unlike the supporters of the proposed allocation, the opponents who tried to defeat the FM shift used both technocratic and nontechnocratic strategies. The three important licensees of Armstrong's system who manufactured most of the FM receivers and transmitters before the war—Zenith Radio, General Electric Company (GE), and Stromberg-Carlson Company—opposed the new allocation. In addition to the three trade associations—representing the television and FM broadcasters as well as the radio manufacturers—the RTPB (especially panel five) also lobbied strongly to keep FM in the lower frequencies. Other opponents included the Yankee Network (a group of FM stations in New England); the *Milwaukee Journal* (which operated WMFM, one of the first FM stations in the country); educational groups who had already invested heavily in FM equipment; and supporters of educational radio (e.g., the U.S. Office of Education and state institutions such as the Michigan Commission on Radio Education). Although the FCC's proposal set aside twenty FM channels for noncommercial, educational radio, educators were worried because some of the supporters of the new proposal continued to criticize the use of nontechnical criteria (for example, the educational value of station programming) in policy making. And—despite Armstrong's claim that the networks were working to defeat FM—NBC and RCA, especially their engineers, also opposed the move. The most vigorous testimony against the allocation during the spring of 1945 came from three sources—Armstrong, panel five of the RTPB, and the FM broadcasters association. Eugene F. McDonald, the president of Zenith Radio, also played an important role in the unsuccessful campaign to defeat the proposal, although he did not testify at the commission's hearings.[43]

Opponents of the FM shift committed to technocratic policies stressed the authority of the engineers who supported their position. The engineers on the planning board were especially upset that the commission seemed to be disregarding the expert testimony of the advisory group it had helped establish. In order to present a united scientific front to the commission, they sought to deemphasize disagreements among the different panels. Opponents of the shift also argued that the vast majority of engineers disagreed with Norton's testimony. McDonald contended that of the fifty-eight witnesses who testified at FCC hearings or voted at the meetings of the RTPB, forty-three recommended keeping FM in the lower frequencies; only eleven approved the shift. Norton, however, did have a significant group of supporters within the engineering community. A few even insisted that informal polls indicated "the majority of scientists agree that FM allocations should be moved upward." On other occasions, engineers responded to McDonald's efforts to quantify or democratize engineering authority by pointing out that "if majority rule prevailed in the field of science, we wouldn't have many inventions."[44]

RTPB engineers acknowledged that since most members were not specialized propagation experts, they were not necessarily the best qualified to evaluate Norton's testimony. The board supplemented the expert testimony they received from Dellinger, chief of the Radio Division of the Bureau of Standards and chairman of the State Department's subcommittee on communications, with testimony from a special committee of six engineers who had "extensive experience in the analysis of data on the ionosphere": Dr. Charles H. Burrows, chairman of the Committee on Propagation of the National Defense Research Committee; Dr.

Harold H. Beverage, associate director of RCA Laboratories and vice president of RCA Communications; Dr. Harlan T. Stetson, director of the Cosmic Terrestrial Research Laboratory of the Massachusetts Institute of Technology; Stuart L. Bailey, a member of the Committee on Radio Wave Propagation of the Institute of Radio Engineers; Dr. Greenleaf W. Pickard; and inventor Armstrong. The group rejected much of Norton's testimony and recommended leaving FM in the lower frequencies. Engineers with the planning board acknowledged that Norton was also an expert on propagation matters, but they stressed that their experts were more qualified because they had a higher standing in the profession. Unlike Norton, they pointed out, "both Dellinger and Beverage have been recipients of the Medal of Honor given by the Institute of Radio Engineers for their outstanding contributions to radio science."[45]

The second technocratic strategy pursued by opponents of the FM shift was to attack directly the technical evidence and reasoning behind the decision. Armstrong and other engineers shared the FCC's public position that such technical decisions as the FM and television allocation should be based purely on engineering evaluation. They wanted to draw a sharp boundary between questions of policy and questions of scientific fact. One of Armstrong's major complaints about the commission was that it "made up the laws of nature to suit itself." Armstrong and other FM supporters believed that the technical evidence was sufficient by itself to justify leaving FM in its original allocation.[46]

In attacking Norton's testimony, they concentrated on identifying errors and mistakes. For example, Norton had argued that in order to determine the strength of F2-layer interference, experts needed to take into account the condition of the ionosphere at places well outside the United States, for instance over the equator. He contended that transmissions from South America would interfere with stations operating in the United States after reflecting off the ionosphere at the equator. Having made this assumption, Norton argued that the extensive data collected at Washington, D.C., by the Bureau of Standards, which did not indicate the possibility of F2-layer transmission above 40 MHz, had no relevance for this kind of propagation. Using data obtained during the war from observations in Hawaii, he argued that the amount of reflection at places near the equator would be much greater than the amount indicated for Washington, D.C. But Armstrong pointed out that transmissions from South America could not arrive in the United States after only one reflection. The equator was approximately three thousand miles from the major population centers in the Eastern United States, and the longest single hop that could have occurred would have been twenty-two hundred miles, eleven hundred miles on each side of the point of reflection. Thus, transmissions from South America could arrive in the United States only after at least two, and probably more, reflections. Because the last reflection point would be within about one thousand miles of Washington, the Bureau of Standards data would be approximately accurate for determination of F2-layer transmission for the major cities of the United States.[47]

Norton admitted his mistake, well before the commission made its final decision, but he continued to maintain that F2-layer interference would be a problem in the 50 MHz band. This would become clear, he predicted, during the next sunspot maximum. Armstrong attacked Norton by arguing that his evaluation was based on highly questionable theoretical predictions rather than hard scientific evidence. He claimed Norton's prediction of the magnitude of the next sunspot maximum and its effects on the ionosphere was "at variance with the history of sunspot cycles during the past 200 years." At a meeting of the Institute of Radio Engineers in 1945, Armstrong referred to his dispute with Norton and the commission as an example of the "age-old battle between theory and practice." During the 1920s, a theoretician had claimed discovery of a mathematical proof demonstrating that frequency-modulation broadcasting would never work. Because of this and similar experiences, Armstrong routinely denounced mathematicians and theoreticians who lacked practical experience. Armstrong preferred Beverage's hard, empirical testimony that the highest observed frequency of F2-layer transmission from Europe or South America had been 45 MHz.[48]

Armstrong also disagreed strongly with Norton about the amount of E-layer interference that engineers might expect in different parts of the 40 to 100 MHz band. He admitted that this kind of interference would likely occur in the lower part of the band, but he believed its effects would be minor, especially compared with tropospheric interference that he claimed would occur in higher frequencies around 100 MHz. In order to undermine the credibility of Norton's testimony, Armstrong and his supporters also pointed out the large number of unstated assumptions that he had used to reach his conclusions. Although this effort to deconstruct Norton's testimony helped clarify important points, it also resulted in a counterproductive round of technical nitpicking.[49]

By themselves, the technocratic strategies pursued by engineers like

Armstrong only underscored the large degree of observational uncertainty and technical flexibility inherent in the effort to allocate FM radio. The testimony of the RTPB's panel five emphasized "the complexity of the phenomena, the interpretations which must be made and the paucity of reliable data." Supporters might have been more successful in convincing the FCC to leave FM in the lower frequencies if they had supplemented the evaluation of technical considerations with nontechnical judgment. Of course, this tactic might also have jeopardized the long-standing relationship that the engineers had cultivated with the commission, which emphasized their special role as pure and unbiased technical advisors. But for the specific controversy about the placement of FM, some recognition of the complexities of this relationship might have helped supporters convince the commissioners to leave FM in the lower band. In fact, a few FM pioneers did pursue nontechnocratic strategies. The owner of WMFM in Milwaukee conceded the importance of engineering considerations, but urged "the commission not to permit them to constitute the sole consideration." He believed that because the engineers "if nothing else" had demonstrated that there was "grave doubt as to the advisability of making the move, . . . the commission's attention should be focused with great emphasis on the nonengineering but otherwise critical factors that are involved." The chairman of the RTPB's panel seven (facsimile) similarly argued that since "no information exists which conclusively demonstrates the superiority of either band," the commission should take into account other considerations, such as the fact that the choice of the lower band would allow "the earlier and more economic production of radio transmitters and receivers for the public." Other supporters of low-band FM insisted that the commission should take into account the need to protect the pioneers of a new public service. This action was necessary, they believed, in order to provide entrepreneurs with "an incentive to invest in new industries."[50]

Despite the recognition by a few participants of the hybrid nature of the decision-making process, the response of the most important opponents of the FM shift was highly technocratic. They might have had a greater chance of success if they had acknowledged the essential nontechnocratic nature of policy making and forced the FCC to take a clear stand on the hybrid relationship. Opponents could have made more effort to pressure the commission to state clearly, for example, the threshold criteria at which point interference could be considered a problem. The government engineers who favored moving FM to the higher frequencies also felt that Dellinger and other experts who opposed the move should have been asked to state clearly "what degree of interference should be tolerated for FM broadcast service." A government engineer wrote Dellinger that "a recommendation as to this policy should have been made by the Radio Technical Planning Board and specific percentages provided for your guidance."[51] Engineering evaluation was important, but it would have been more effective if it had been used to convince the FCC to acknowledge that other considerations also needed to be taken into account. Opponents of the move could have combined this strategy with an effort to point out inconsistencies in the commission's technocratic-legitimation strategy. Although decisions were justified based on engineering expertise, the public record clearly indicated that the commission had also considered nontechnical criteria.

If Armstrong and his allies had spent less time pursuing technocratic strategies, they might have done a better job providing the FCC with a clear view of the economic and social effects of the proposed allocation. For example, they could have more effectively testified about the amount of time the industry needed to convert to the higher frequencies, partly by actively refuting alternative testimony. One contemporary observer believed that "one of the . . . factors which prompted the Federal Communications Commission to allocate basically on engineering considerations was understood to have been [the] refusal of manufacturers to state definitely that they would turn out sets with a 2–1 rejection ratio." The two-to-one rejection ratio referred to FM receivers capable of discriminating between two signals until the weaker signal was half as strong as the main signal, at which point interference would occur. The FCC engineers predicted a large amount of sky-wave interference, partly because they assumed a ten-to-one rejection ratio for receivers.[52] Had the FM manufacturers done a better job publicizing the quality of their product, the commission might have seen sky-wave interference as a minor problem.

After 1945: New Strategies

During the controversy over the new FM allocation, the low-band FM supporters generally avoided starting a public relations and lobbying campaign. They did not think it was proper to debate publicly issues that they believed were fundamentally engineering in nature. But after the FCC rejected a final request, in January 1946, to salvage the old FM system by allowing FM broadcasters to use both the old and new bands, the

Armstrong forces changed tactics. On January 30, an observer reported that the proponents of low-band FM "have cast aside the kid gloves with which they have been sparring and are now going in for slugging." Beginning in 1945 and especially during 1946, Armstrong and the FM supporters gained a better appreciation of the need for political, business, PR, and organizational strategies in order to demonstrate the superiority of FM radio and achieve commercial success.[53]

Armstrong and the FM pioneers partly felt that a new approach to the FM controversy was needed to counteract the "careful and premeditated public relations" campaign being conducted by radio and television manufacturers and broadcasters. This was especially true during the late 1940s, after Armstrong filed suit against RCA for infringement of his FM patents. But as early as March 1946, Armstrong hired the services of a PR firm to help publicize and promote FM. Armstrong gave his PR consultants general ideas about what to write in articles and papers, and the consultants submitted publications in Armstrong's name. They sent "releases" promoting FM and criticizing the policies of the FCC to numerous newspapers and magazines. Armstrong and his consultants worked especially closely with trade publications such as *Radio and Television Retailing* and *FM and Television.* The consultants sent Armstrong copies of stories about FM from the same publications. Armstrong paid his consulting firm $1,000 per month during 1947. He also personally promoted his invention by giving speeches, writing letters and articles, and participating in publicity stunts. In 1947, for example, Armstrong agreed to attend a "publicity stunt" sponsored by a Chicago FM station, which attempted to demonstrate, in an entertaining way, the capabilities of FM radio to 450 members of the studio audience that included the "*top* dealers and salesmen in the Chicago area, carefully selected through the business survey facilities of the *Chicago Tribune.*"[54]

The PR consultants also tried to coordinate their own "FM public relations" with similar marketing and advertising activities being carried out by FM trade associations and FM manufacturers. In 1946, one of Armstrong's PR consultants wrote McDonald, the president of Zenith Radio, that "if there was something we *all* could combine to do in the way of a long-range program of public relations, something that most of those deeply interested in FM could get behind, we might find that FM's progress could be accelerated." During the late 1940s, the FM pioneers worked collectively not only to try to convince the public to use FM but also to encourage manufacturers to produce a large supply of FM receivers and to "effect national recognition of FM's virility as an advertising medium." One of Armstrong's special concerns was to make sure that radio dealers avoided cheap imitations by selling only receivers licensed under his name.[55]

Political lobbying, too, became important. As early as spring 1945, McDonald convinced the other pioneer FM manufacturers (and Armstrong licensees) to contribute to a letter-writing campaign encouraging members of Congress to pressure the FCC to support low-band FM. Both McDonald and Armstrong intensified their lobbying during 1946. Armstrong was a significant contributor to the Republican Party, especially to the campaign of Senator Charles Tobey of New Hampshire, who subsequently became one of his strongest allies in Congress. Armstrong and McDonald also worked to enlist the support of the most important member of Congress involved in radio regulation, Senator Burton K. Wheeler, chairman of the Senate Interstate Commerce Committee. In January 1946, McDonald assured Armstrong that "tonight at dinner, I will make it a point to have a talk with Senator Wheeler on the subject of your accomplishments. I do not think he realized what you had contributed to radio any more than did Senator Tobey until he was given the details."[56]

At first the two men were unsuccessful in getting Congress involved in the dispute over the FM allocation. Many congressmen were hesitant to confront a problem that members of the commission convinced them was fundamentally technical rather than political. But not all members of Congress shared this view about the role of politicians in technical policy making. One senator argued that the "very idea of complexity and confusion and technical abstruseness has been sown in the Congress and spread deliberately both within and outside the commission to shut out prying minds." At first, during 1945, a few congressmen responded to McDonald's campaign by writing the FCC in support of low-band FM. After three years of intense lobbying, Armstrong and McDonald convinced the Interstate and Foreign Commerce Committee of the House of Representatives to hold an official hearing to consider a resolution to assign the 50 MHz band to FM immediately. But enough members to defeat the resolution considered the proposal improper interference in the technical work of an independent regulatory commission. Armstrong admitted that he did not like the idea of legislating allocations, but he believed that "we are faced here with a condition that we must have relief from somewhere."[57]

In their effort to establish the technical superiority of FM radio, Armstrong and his allies attempted to enhance the effectiveness of their lobbying and popularizing campaign by emphasizing particular political, social, and cultural meanings for FM. Most important, FM was characterized in utopian terms as a new technology that would solve the major problems resulting from the development of AM radio. Supporters argued that, unlike AM, FM was inherently democratic and would free broadcasting from both corporate domination and government control. A speech written for Armstrong by one of his PR consultants argued, in technocratic terms, that "just as technological factors of scarcity have made this censorship possible, just so has technology come to the rescue by undoing the scarcity and indirectly exposing the censorship and monopoly which it supported." FM was characterized as the "biggest single advance in the history of radio" and "the great boon to public service, democratic communication and cultural enlightenment." Although only a few stations could broadcast on AM channels without causing interference, FM channels could accommodate hundreds of stations. McDonald promised congressmen that "FM can provide an interference-free station in every city in the U.S. over 2500 population." FM supporters believed that the shortage of AM channels had been the most important factor supporting the growth of network monopolies. They also contended that the FCC opposed FM radio because the new system threatened to undermine government regulators' domination of broadcasting.[58]

The argument for the inherent democratic character of FM radio helped convince key social groups, especially agricultural organizations, to lobby Congress to authorize low-band FM. These groups were convinced that low-band FM was especially important because its propagation properties were more conducive to long-distance rural coverage. During the weeks prior to the 1948 congressional hearing, more than two hundred individuals and seventy-five organizations mainly representing agricultural interests sent letters to members of Congress urging them to pressure the FCC to authorize low-band FM. But in 1947, the FCC had assigned the 50 MHz band to emergency and mobile services. Thus, numerous police organizations, fire departments, state forestry bureaus, and highway departments lobbied against the proposal.[59] What many participants had originally viewed as primarily a technical problem became—thanks partly to the work of Armstrong and the FM pioneers—a highly political controversy.

This chapter gives us a deeper understanding of FM radio's "failure" to live up to the initial expectations of its early enthusiasts. Proponents of FM correctly identified decisions and actions that were made by some of the large manufacturers and broadcasters against FM (such as the decision to commit limited resources to developing television instead of FM), but the narrative these FM champions created tended to collapse the entire complex story of FM development into a simpler history of the "individual warrior struggling against organized evil." Complexities, such as the role of the two competing television systems in the 1945 FM allocation decision and the fundamental disagreements among commissioners and staff, were played down or ignored. Also, Armstrong's historical narrative failed to acknowledge that the 1945 decision actually seemed to favor FM by authorizing more channels than the FM industry itself had requested (at least ninety, compared with seventy-five). Further, although Armstrong's supporters argued that the FCC's allocation decisions in the VHF band were motivated by a desire to help the television industry, during the period from the late 1930s to 1946, FM actually *gained* channels in the VHF band at television's expense: while the number of effective FM channels increased from fewer than thirty-five to ninety, the television assignment decreased from nineteen channels to thirteen. Moreover, when discussing the inherent technical superiority of FM, Armstrong tended to ignore the fact that some of FM's advantages—especially its capacity to eliminate static and the ability for hundreds of stations to operate in each channel—partly resulted from the unique propagation properties of the higher frequencies in which it operated. If officials had allowed AM stations to broadcast in the VHF spectrum, those stations would have enjoyed some, though not all, of the same advantages.[60]

A complete explanation of the "failure" of FM would need to take into account other historical developments, including the rise of the television industry during the 1950s and the impact of other decisions by the FCC after World War II. What is of more importance for this chapter, however, is that we avoid taking an uncritical, teleological view of technological development. Rather than assume unproblematically the inherent "technical superiority" of such inventions as FM radio and look for grand conspiracies to explain their suppression, historians need to take into account the complex nature of regulatory decision making, the defining role of different institutions and individuals, the contingencies of historical context, and the essential role of nontechnocratic strategies in shaping technological development.[61]

5)))

VHF and UHF

Establishing a Nationwide Television System, 1945–1960

It is my belief that it would be just as criminal to hold back television as it would be for a scientist to keep from the public a known cure for one of mankind's great ills, once he had discovered it. The analogy is fair, for Television will bring about the enlightenment of mankind, and may well hold within its grasp the solution of a lasting peace for the world.

Norman D. Waters
of the American Television Society,
October 12, 1944

Planning for postwar broadcasting began as hostilities in Europe were winding down. The competition between FM and television for prime channels in the spectrum was only one conflict among services seeking to use VHF frequencies. Because of this competition, the FCC gave only twelve channels to television in 1945, six fewer than the prewar authorization and many fewer than most industry officials thought were necessary for a nationwide competitive system. After deciding against immediately authorizing new commercial channels in the UHF using color or high-definition standards, the commission went ahead with an assignment plan for television stations using the limited number of VHF channels. However, in 1948 planners questioned their earlier assumptions, especially the estimates of potential sources of interference.

During the television "freeze" of 1948–52, the FCC refused to consider new applications, pending a review of assignment policies. A revised assignment plan for television stations, which used approximately seventy newly authorized UHF channels, was put into place at the end of the freeze. But by the middle 1950s, it became clear that the commission's original intention of establishing a nationwide system based on a policy of having stations in particular markets compete on an equal level had not been realized. Most important, UHF stations could not compete with the better-established VHF facilities. While the number of VHF stations increased from approximately 250 at the beginning of 1954 to more than 440 by the end of the decade, the number of UHF stations dropped from more than 125 to fewer than 80. And although the commission assigned television channels to more than 1,250 communities at the end of the freeze, by the summer of 1958 television service existed in only approximately 300 cities. This overall trend helped support the dominant position of the two networks, NBC and CBS, which controlled network access to most of the VHF stations in the largest markets. Newer networks, notably ABC and DuMont, had difficulty competing using the available UHF outlets.[1]

In chapters 2 and 3, we saw how the allocation and assignment decisions for AM radio played a crucial role in helping to establish the "American system" of network commercial broadcasting. The establishment of an assignment plan for television stations was also based on this model, but its particular character depended on the key commission decisions of the postwar period. Important questions partly raised in earlier chapters that we need to consider here include: To what extent did the technocratic perspective so important in the case of AM radio also play an important role in the effort to establish a national television system? How did the participants deal with the relationship between the "technical" and "socioeconomic" aspects of television allocation and assignment? What arrangement did the commission use to institutionalize the advisory role of engineers? How did the commission deal with the tension between advocacy and objectivity—that is, the problem of relying on technical testimony from experts who were increasingly likely to be employed by large corporations with an economic interest in the developments under review? Finally, how did the participants address the problem of having to plan industry development based on technical knowledge that was in a number of cases highly uncertain?

Competition for Channels, 1944–1945

Before the FCC developed a complete assignment plan for television stations for the postwar period, it needed to make a final decision on the allocation of channels. Competition with FM accounted for only part of the reduction of television frequencies in the VHF. Other new services, too, sought to expand operations in the limited band of VHF frequencies after the war. These new users included power companies, utilities, forestry and conservation agencies, railroads, taxi and bus companies, highway departments, common-carrier operators, and emergency services such as fire and police departments. But perhaps more important were the new uses for radio in support of aviation. During the commission's allocation hearings in 1944, T. A. M. Craven, an engineer who had served on the commission before the war, contended that "two great industries are in competition for portions of the radio spectrum between 30 and 1000 mcs. These industries are aviation and radio broadcasting [mainly television]. . . . Both industries will have a profound effect upon the social and economic structure of the nation. Both are needed by the country to help mitigate post-war economic problems." Panel eleven of the Radio Technical Planning Board (RTPB), the industry advisory group responsible for evaluating the needs of aeronautical radio, claimed that "it is no exaggeration to state that without radio, air transportation could not exist, at least to any extent substantially contributing to our civilization."[2]

Developments in military aviation during the war helped drive the growth of the commercial aeronautical industry. The military's expanded interest in radio, especially in aviation, continued into the postwar period and helps account for the reduction in channels available to television. The wartime trend is evident in the changing use of the spectrum. In 1939, 26 percent of the frequencies below 162 MHz were allocated exclusively to government agencies, including the military; by 1943, the number had increased to 37 percent. During this same period, the percentage of government stations sharing frequencies with nongovernment users of radio increased from 8 percent to 32 percent. Above 162 MHz, the U.S. Army and Navy had exclusive control of 47 percent of frequencies in 1943 and nonexclusive use of 48 percent. Also during that same year, the military demands on radio led the Interdepartment Radio Advisory Committee (IRAC), the government organization in charge of assigning government stations, to begin to deny requests from nonmilitary federal agencies for radio facilities. During the World War II emergency, the War Department applied directly to the FCC for at least one of the channels that the commission had set aside exclusively for television. An official with the commission warned that this would be a "dangerous precedent"; he feared that the military might resist moving their transmissions back to government bands after the war ended. Different groups, including police and other emergency services as well as television broadcasters, especially coveted the band of frequencies between 108 and 132 MHz, which military aviation used during the war. The chairman of the police committee of the RTPB's panel thirteen feared that "by the time the war is over aviation may have so much equipment in operation, with hundreds of big military planes converted to commercial 'freighters' and passenger planes, that they will not want to change."[3]

When members of IRAC began to plan for postwar use of the spectrum, they assumed that the military services would not return to a prewar level. Since the military use of radio was largely classified, they were forced to accept without question the military's recommendations for frequencies in the postwar period. Craven, who became chairman of the committee in 1943, admitted that he "didn't press too strongly to get the details. I was very much influenced by the experiences of the military departments and accepted without question their judgment in these special matters." Craven further pointed out that anyone wanting to operate in portions of the spectrum being used by the military would essentially be asking them to give up equipment set up for those frequencies and funded with "a billion dollars of the taxpayers' money."[4] This was the situation facing VHF television during the postwar period.

The competition among different services for channels was apparent in the conflicting allocation proposals developed by the interdepartment committee and the panels of the planning board. In their efforts to reconcile the conflicting demands of users of the spectrum, members of these groups had to deal with the central problem already discussed in the case of the FM allocation—the relationship between technical and nontechnical factors in decision making. In general, they first assumed the existence of a sharp boundary between the two kinds of considerations and tried to determine an ideal "scientific" allocation that would reflect the best technical locations for different services in the spectrum, taking into account objective physical constraints (e.g., the propagation properties of different regions of the spectrum and interference that might result from different services operating on adjacent channels). At one of the first meetings of IRAC to consider postwar planning in the use of the spectrum, Craven announced that "in the beginning we will consider it on a scientific basis and determine what the principles of allocation should be. . . . Later, we will consider the practical situation and see if we can fit it in." Officials asked individuals testifying for different services at these meetings if the "technical considerations are such that it definitely tells us where it belongs." Craven felt that they should not take into account the "economics of the investment in the situation" because "then you get a hopeless compromise between engineers."[5]

But as in the FM allocation, the engineers could not always agree about the technical problems. RCA engineer Charles Jolliffe did not believe there were any "scientific 'musts' for placing individual services in particular locations in the frequency spectrum." On close inspection, arguments that engineers claimed were based on "technical" factors also seemed to involve economic considerations. For instance, representatives of the television industry argued that the "technical reason" for needing frequencies below 250 MHz was because engineers had not yet developed the proper equipment for the higher bands. But their main concern seemed to be an economic one—that "if television does not get the channels in this range there will be no television for some time." Although they were not always consistent in their views, other engineers, including Alfred Goldsmith of RCA, asserted that "engineering takes into account the dollar sign and, therefore, is not pure science." As we saw in the preceding chapter, W. R. G. Baker took an even stronger stand when he declared that "I don't differentiate between the economic and the purely technical." He questioned whether any "natural" place in the spectrum existed for television or most other services. And since any engineering evaluation of the allocation problem would generally involve economic considerations, he concluded somewhat flippantly that "if there are 1000 engineers, you will have 2000 opinions."[6] The different uses of the term *technical* contributed to a blurring of the boundaries between different kinds of considerations. The main use referred to certain constraints or limitations imposed by physical attributes of the natural world, but this physical meaning was also used to refer to definite constraints imposed by man-made equipment. The relationships among different considerations used by engineers advising the commission were further complicated by the fact that policy decisions normally had to be made based on uncertain and limited knowledge.

Partly because these problems were not always clearly dealt with, a critical view of the role of engineers as policy advisers became more prevalent during the postwar period. Edwin H. Armstrong, who felt slighted by the commission for policies that seemed to stifle the development of FM radio (see chapter 4), accused the engineers on the commission of taking "positions of advocacy" and then attempting "to establish that the scientific facts, the laws of nature supported these positions." He was not alone in such views. Craven complained that "if we wait upon the scientists to decide upon standards, we will never make a decision. These decisions always were and will have to continue to be made by administrators." Commissioner Fly criticized industry engineers "who put out policy conclusions under the cloak of technical observations." Despite such comments, it was not unusual for an engineer to testify that he was "speaking solely as a scientist, not as a representative of any individual or any group. I have no personal ax to grind, and I don't represent anyone who has."[7]

But the critical view was also reflected in the public press and in congressional testimony, especially as the commission made controversial decisions about color television and UHF broadcasting and as industry engineers testified as both members of advisory groups and representatives of manufacturers or broadcasters. A 1946 article in *Fortune* magazine complained that "it is nothing to hear one imposing engineer say that no tube is available to give sufficient power output at high frequencies, while another imposing engineer says he has one and it does. It is nothing to hear one engineering group say that the wide future of television is in the ultra-high frequencies, while another swears that, though it's feasible, there's actually no practical advantage up there."[8]

At least compared with the earlier periods when Hoover and the radio commission had authority over the regulation of broadcasting, by the late 1940s and 1950s the FCC was more likely to acknowledge publicly that its decisions involved not only technical but also economic and social considerations. For example, the commission acknowledged in 1945 that its decisions about allocating services to different parts of the spectrum had not been "limited to technical considerations but also took into account economic and social factors and considerations of national policy."[9] This position differed from the justification for the decision about the placement of FM because FM was not seen as a service different from television and other forms of broadcasting. And unlike during the 1920s, when members of the Institute of Radio Engineers, a professional engineering society, played the most important advisory role, during the 1940s and 1950s engineers organized by trade associations like the Radio Manufacturers Association played an equally important part in advising the government. It seemed appropriate to use technical experts from the manufacturers association because their engineering analyses would reflect both the technical and economic concerns of the industry.

To supplement the "technical" or "scientific" evaluations that might help determine which users would receive access to different parts of the spectrum, industry and government planning committees—especially IRAC and the FCC—established "an order of priority" for the different services. In its recommendations for industry use of the spectrum, the radio planning board also recognized that officials would have to make compromises because the demand for channels exceeded the supply available; however, because its charter formally limited it to technical evaluation, it did not explicitly establish similar rules. The hierarchy first established by the interdepartment committee in 1944 and partly adopted by the commission during the following year gave first priority to the "preservation of life and property where other means of communication are not available." This decision was based on a tradition going back to the earliest efforts to regulate radio. Secondary status went to "essential communication services which must use radio because no other method of communication can be used." This included aeronautical and maritime stations, which, unlike many operations over land, could not take advantage of cables. Officials delegated radio broadcasting, including voice and television transmission, to the next order of priority. The commission took into account these same priorities but also used other criteria to help make decisions, including giving consideration to services benefiting large groups, asking if the public would likely accept a new service and, when determining rival requests, taking into account the economic investment in equipment.[10]

The priorities established by the interdepartment committee were particularly important in blocking expansion of television well into the frequencies above 100 MHz. Craven pointed out during commission hearings that both commercial and military aviation needed radio for reasons having to do with safety. Broadcasting, by contrast, required radio "for social, cultural, educational, and entertainment purposes." Further, while television had made little investment in transmitters or receivers using frequencies above 100 MHz, aviation had invested heavily and would likely need more space in the future. Other mobile and fixed services requesting the same channels as television, including police and emergency radio, also clearly operated for reasons of public safety. One group that had difficulty competing with either television or the commercial and government services was that of the amateurs. This was a continuation of the trend that saw amateur operators lose influence in policy debates. When an amateur representative suggested that television could make do with fewer VHF channels because it was a luxury rather than a necessity, he was quickly placed on the defensive and forced to state that "the last thing I want to happen is for anybody to think that I am an enemy of television."[11]

Promoters of television wanted more VHF channels than the eighteen assigned in 1941; their ideal was thirty to forty channels. Panel six of the planning board recommended twenty-six; panel two, which was responsible for reconciling conflicting demands of different services, reduced this recommendation to eighteen. Citing the government's need for the channels between 40 and 250 MHz, the interdepartment committee recommended television be given fifteen channels. A request for the ideal of thirty channels, the committee pointed out, would monopolize 85 percent of this band of frequencies.[12]

The reasoning behind officials considering thirty channels to be an ideal again underscores the important tension between technical and socioeconomic evaluation or legitimation. Interference between stations too close together was an important technical limitation that needed to be taken into account when the commission assigned stations to cities. Information about this form of interference was limited, especially for stations using frequencies above 150 MHz. Planners also favored a large number of stations to create a competitive industry offering a variety of programs. The major northeastern cities would have considerable markets that could support more stations, but because these broadcasters would also be located close together, they would be especially sensitive to interference. Planners had to balance these technical and economic issues with social and political decisions about whether small cities should be allowed to have their own local stations. By taking into account co-channel and adjacent channel interference and by providing multiple service to the large urban areas along the Atlantic seaboard, panel six of the planning board calculated that thirty channels would allow eleven stations in Philadelphia, two in Baltimore, two in Providence, and fifteen in New York. However, as in the case of policy making for FM, the planning board argued publicly that its recommendation was simply based on a technical evaluation: "While the panel was not in a position to consider the economics of the situation, it felt that the technical considerations which have just been discussed indicated that approximately thirty channels will be desirable to provide a nation-wide competitive television broadcasting service." A recognition of the inconsistencies between this statement and previous testimony led one of the members of the commission to point out ironically that "without considering the economics of the situation, the panel has secured the answer to an economic problem."[13]

Because of pressure from other services for use of the VHF band, the commission's January 1945 decision on the allocation of the spectrum reduced the number of channels available for television to twelve, of which eleven would have to be shared with government and nongovernment stations, fixed and mobile, in regions where interference would not be a problem. The FCC acknowledged that this would not provide enough channels for a "truly nation-wide competitive service," but the members felt that it would provide sufficient service until television broadcasting was ready to move to the UHF band it set aside for experimental television broadcasting.[14]

VHF versus UHF: Negotiating Technical Uncertainty

In the preceding chapter, we saw how CBS had unsuccessfully pushed for immediate transfer of television to the UHF. This effort was closely tied to its attempt to introduce a color-television system and gain an economic advantage over RCA through control of the key patents for this rival system. The company also argued that UHF would provide enough room for a higher definition black-and-white system. CBS officials contended that this would be the last opportunity to upgrade television standards before they became frozen once the industry had sold millions of receivers. If the commission had decided to upgrade in 1945, only about 10,000 prewar receivers would have become obsolete. The commission's decision to shift the FM band, which forced consumers to buy new FM receivers, served as a precedent for CBS's proposal. A much earlier precedent was the conversion of spark transmitters on ships to continuous-wave apparatus. In this case, the government had provided a transition period during which it allowed both types of apparatus to operate.[15] The FCC might have made a similar decision in 1945 by authorizing commercial transmission of both monochrome VHF and either monochrome or color UHF (as soon as it was ready), but this action would not necessarily have kept the old apparatus from using its market advantage to freeze out the new system. The debate over color remained tied to the controversy over UHF until the late 1940s, when engineers discovered how to fit their color systems into the narrow-band VHF channels.

Despite chairman Fly's support in 1944, many of the large and well-established members of the industry—including most importantly RCA and NBC, but also Philco, Farnsworth, DuMont, Don Lee Broadcasting, and to a less extent GE—opposed CBS's efforts. DuMont had been one of the most vocal opponents of the RCA-inspired standards of the National Television System Committee before the war, but was now ready to profit from their adoption. The newly formed Television Broadcasters Association also refused to support CBS. RCA, the broadcasters association, and CBS's other opponents wanted immediately to resume operations in the VHF band after the war, using the prewar standards in which they had already invested substantial resources. New standards would require manufacturers to retool factories, which would delay production and prevent them from taking advantage of a pent up demand for consumer electronics. Labor unions, as well as a small, nonprofit organization called the American Television Society, supported the RCA "television now" campaign because it promised to create jobs as soon as the war ended. Arguing that "it served no master but the public interest," the American Television Society used utopian language to emphasize that since "television will bring about the enlightenment of mankind" the government should not delay its introduction. Cowles Broadcasting and—slightly less enthusiastically—Zenith, Westinghouse, and the Federal Telephone and Radio Corporation supported the CBS position—the last three companies by offering to build transmitters and other equipment for experimental UHF broadcasts. Fly's departure from the FCC in 1944, as a result of congressional and industry opposition to his controversial positions, left CBS with little support on the commission. But the commission did hold out some hope that broadcasters could eventually use the experimental UHF frequencies for a new system.[16]

The resolution of the debate over color and UHF depended on how individuals dealt with the problem of making policy decisions based on technical knowledge that was not entirely certain. CBS's proposed color system needed further development; no workable system existed in 1944 to demonstrate to policy makers. CBS argued that a relatively short period of concentrated effort would solve the remaining problems; a committee of the RTPB, by contrast, thought it would take another five to ten years.[17] Further, experts had little understanding of the propagation properties of frequencies in the UHF band, and the industry still needed to develop tubes able to transmit the high-power signals needed for television broadcasting.

Engineers and scientists made progress during the war with both tube development and investigations of propagation characteristics, but the classified nature of this information complicated the efforts of policy makers. CBS played up the technical advances of the war and argued that high-power UHF transmitters would soon be ready. CBS television engineer Peter Goldmark testified in January 1944 to an IRAC subcommittee on postwar planning that "from the day the war stopped and secrecy was lifted from certain developments, I'm certain within one year from that date a suitable television transmitter in that band could be produced." It should be kept in mind that in 1944 officials widely believed that the war would not end until 1946. Other engineers, notably those employed by RCA, believed the industry would need more than one year to develop the proper equipment once the war ended. They also pointed out that the industry would need to accomplish much more than develop tubes for the high frequencies. According to RCA chief engineer Jolliffe, an "entire system must be proven." Jolliffe listed the steps involved in the process of authorizing a new system of commercial broadcasting: first, broadcasters would need to install experimental transmitters and receivers; second, the industry would have to develop and test engineering standards; third, companies would need to evaluate public reaction and make changes to incorporate these results; fourth, the FCC would need to review the system recommended by industry; and finally, after the commission authorized commercial development, manufacturers would need to gear up for production. This entire process, according to Jolliffe, would take more than five years.[18]

In their attempt to formulate policy based on uncertain and limited information, engineers pointed to lessons drawn from history. Some experts reminded government officials that the lack of knowledge about the AM broadcast band during the 1920s had not stopped the Department of Commerce and industry from making important policy decisions. Jolliffe used this argument to convince the commission to authorize VHF television, despite the fears that the limited number of channels would not provide a nationwide interference-free service. In a letter to chairman (and former chief engineer) Jett in December 1944, he urged "somebody in authority . . . to crack heads together and say that things must be made to work and not object to taking reasonable engineering risks." According to Jolliffe, "no allocation problem ever was solved by going out and finding all the answers and meeting all the contingencies in the field in advance. If we had tried to do this in broadcasting, in high frequency communication, in aviation, etc., we would be many years behind our present position." But CBS also used this idea of taking "risks in order to obtain progress" to support its belief that the commission should authorize the commercial exploitation of the UHF band despite the lack of knowledge about its use. Furthermore, at least one engineer rejected Jolliffe's arguments by pointing out that many of the problems with AM radio, such as the inadequacy of the assigned band of channels to meet the huge demand, might not have occurred if officials had had a better understanding of both the propagation characteristics of the frequencies and the social and economic implications of the new technology.[19]

One of the key disagreements among engineers debating the future of television was the amount of technical progress made during the war. GE engineer Baker, who generally tried to remain neutral on the CBS proposal, testified during a meeting of the interdepartment radio committee that "one of the confusing parts of this whole business is the overstressing and over-selling of the technical accomplishments due to the war. . . . There has been too much bally-hoo on that in my opinion." By contrast, several engineers working for the military, including one on leave from NBC, supported CBS's contention that war research demonstrated the feasibility of moving television to the UHF band.[20]

Despite this military testimony, the FCC did not think experts knew enough about UHF to justify authorization of commercial operations in 1945. The commission assumed that such a move would cause a major delay in the resumption of television after the war and any delay would not be viewed favorably by the public or by broadcasters and manufacturers.[21] Commissioner Jett, who became FCC chairman after Fly's departure, was not inclined to take a strong stand to interfere with industry's development. He had opposed Fly's New Deal efforts and now had an opportunity to allow the companies that had invested in standards developed by the system committee to continue commercial development.

Compromising on Assignments for a Limited Band

After the commission decided in 1945 to authorize commercial VHF television, the next major task was to establish a nationwide assignment plan for stations. The FCC wanted to avoid the same interference problems that had occurred when AM stations were established haphazardly on a first-come, first-served basis. Government officials hoped that by determining where broadcasters should locate stations and the number of stations that a given market could sustain without creating interference, the assignment plan would simplify and rationalize the FCC's task of evaluating the proposals of individual applicants. The commission gave television a thirteenth channel in May, but most officials believed this would still not support a competitive national system. With only thirteen channels (six below 108 MHz and seven above 174 MHz), government planners would not be able to realize their stated goal that "broadcasting service should be available to all communities, whether it be sound broadcasting or television." The commission pointed out that "it would be impossible to take care of cities like Paterson, N.J., White Plains, N.Y., and so on."[22] Nevertheless, after the May decision, the staff—in consultation with industry—began to consider how to establish an assignment plan for the entire country.

The interest in developing postwar television as rapidly as possible meant the FCC and the industry were very willing to make compromises, including following Jolliffe's advice on "taking reasonable engineering risks." Instead of consistently emphasizing that the VHF channels were only temporary and would be discontinued once UHF was ready for commercial broadcasting, chairman Jett told broadcasters that "I do not want anybody to take that phrase 'temporary character' too seriously." At a press conference in January 1945, he informed the audience that "I wouldn't want you to assume that I mean that when we do go forward with the ultra-high frequencies . . . that the lower channels would be discontinued. I personally have no such thought in mind." The representative of the Television Broadcasters Association announced that "'temporary assignment,' as far as the industry is concerned, means nothing."[23]

In their eagerness to push ahead with VHF television, industry supporters also played down the inadequacy of the band, contending that thirteen channels would in fact provide good service. Philco pointed out that it would "enable several hundred stations to go on the air after the war, and give a large portion of the public a regular television program service." Even twelve channels, according to the Television Broadcasters Association, were "capable of providing a competitive nation-wide service to a majority of the people of the USA." The industry representatives also emphasized that either twelve or thirteen channels would provide sufficient service to open up many job opportunities after the war.[24]

The commission was generally willing to compromise by adopting some of the key recommendations of the television association. Since the association mainly represented the well-established broadcasters based in the largest cities, the assignment proposal, not surprisingly, gave the most channels to the largest cities. Significantly, the association recommended that New York City should receive seven channels, the maximum number possible for one city in order to avoid interference between any two adjacent channels of the thirteen total.

An Industry-Commission Television Allocation Committee at first rejected this proposal during meetings in July and August. The committee was made up of engineers representing the FCC, the War Department, Philco, DuMont, RCA, NBC, and television station WOR in New York City. At the first meeting, the chairman instructed them to take into consideration not only technical factors such as sources of interference but also "economic factors," notably the need to create competition within a given market. Instead of seven, the committee recommended only four channels for New York City. It based this decision on the FCC's desire to allow smaller cities, especially in the Northeast, to have their own stations. Instead of primarily taking into account market conditions, the committee wanted to base the assignment plan on "population and distribution areas." Some of the commissioners thought this was necessary, partly because of key statutes of the Communications Act, which required the commission to distribute broadcast facilities "on an equitable basis" to all regions of the country. Chairman Jett testified that "I do not see how we could give seven channels to New York and deprive the other cities . . . of television service without violating the Communications Act."[25]

During fall 1945, the television association continued to try to convince the commission to authorize seven channels in New York City. The organization criticized the FCC for relying on engineers who were not entirely qualified to take into account the practical economic realities. "While the question of interference is one of the controlling factors," according to the association, "there are other factors which should also be given consideration. These are economic factors and factors of public service." Specifically, the major broadcasters believed the engineers did not sufficiently appreciate the importance of New York City as the preeminent competitive market where economic development would begin.[26]

Since the FCC did not find this argument persuasive, the television association proposed a technical solution to the problem. Stations should use directional antennas to reduce interference. But the needs of aviation again blocked this potential development for television. The FCC rejected the proposal based on the belief that "with the great increase in civil aviation as a result of the war, it is going to be increasingly difficult to find suitable antenna sites that do not constitute a hazard to air navigation."[27]

Another nontechnocratic proposal drew on the record of AM radio during the 1920s and 1930s and raised fundamental social, political, and economic questions about the place of broadcasting in U.S. society. A number of small broadcasters and the newest network, ABC, suggested that the commission allow two or more television stations in a particular city to share a channel, and even a transmitter. Not only would this provide television service to most of the country by allowing more stations to operate, but it would help prevent monopoly control of a "perfect media for propaganda," an issue that remained a significant worry for a number of Americans. The engineer and president of American Television Laboratories, U. A. Sanabria, had been an outspoken critic of the FCC before the war for not taking a strong stand against the dominance of television by big business, especially RCA and the networks. During the postwar debate, he again insisted that the commission consider the monopoly issue when determining an assignment plan for television. Instead of trying to find technical solutions to the problem of allocating and assigning channels, Sanabria argued that "it is time we slice this technical fog and stop thinking of how more stations may be introduced without first democratically distributing the ones we know can be used." The FCC had attempted to prevent monopoly control by proposing a five-station limit on the number of television stations a company could own, but Sanabria believed "it is how you manage to distribute those five key channels that makes for no monopoly." Sanabria proposed that the first channel authorized for a large city go to "big business applicants" who would share time on a master transmitter; the second and third channels would go to "collective small business applicants," composed of as many as seven stations organized as a nonprofit corporation and sharing equally in the ownership of a powerful transmitter. Each operator would transmit during one day every week; since more money could go into programming rather than equipment, superior broadcasts might result. The fourth and fifth channels would be distributed to labor, educational, and religious groups, and to stations owned by the theater and movie industry. Sanabria believed the government should always give preference to "collective ownerships which . . . better represent a cross section of the business of our United States."[28]

The FCC seriously considered the channel-sharing proposals, but chose not to adopt them. With the departure of Fly, the commission was less likely to take a strong stand against the large broadcasters and networks. Both CBS and NBC as well as the trade associations representing the television broadcasters and the radio manufacturers opposed channel sharing. Opponents correctly pointed out that the policy had not been very successful when the government used it for AM radio. They reminded the commission that it had been "a disturbing factor leading to constant argument and bickering among licensees." Opponents also contended that rather than help stations economically, channel sharing would make it difficult for stations to survive and prosper. If they could not operate full-time, opponents believed it would be difficult to attract "adventurous capital necessary to develop the television field."[29]

One of the best examples of the FCC's willingness to compromise with industry and take "reasonable engineering risks" was the approach to the problem of tropospheric interference. The FCC's underestimation of the importance of this form of interference was a major reason for the freeze on new stations and the reevaluation of policy in 1948. Officials needed to take into account all sources of interference between television signals before they developed an assignment plan. Like F2 and sporadic E interference in the lowest part of the VHF band, tropospheric interference throughout the band was suspected of being a problem, but the knowledge was limited and uncertain. Tropospheric interference occurred when signals not only propagated normally as ground waves but also through scattering in the atmosphere because of discontinuities in temperature and density.

During summer 1945, the Industry-Commission Television Allocation Committee recommended that the FCC separate high-power television stations operating on the same channel by 170 miles, and high-power stations operating on adjacent channels by 85 miles. There seems to have been much uncertainty during the meetings about how to handle tropospheric interference. Some witnesses presented testimony during the industry and FCC hearings in 1944 and early 1945 indicating that this form of interference might be a major problem. One engineer testified that "tropospheric interference is certainly an unknown and of possible serious trouble. We do not know what to expect." Another warned that "our present knowledge of tropospheric effects, does not extend over much of the band under consideration. . . . To protect stations distance of separation may need to be doubled." But apparently because of uncertainty about its effects, the committee did not make major adjustments to their recommendations. The committee asserted that "it was found most important to consider tropospheric data, but such a subject was found to be difficult to handle." After the July 19 meeting, Goldsmith reported that "the committee would present to the commission the best plan possible at this time but recognize the fact that there were certain shortcomings in the plan which did not take into consideration tropospheric" interference. The RTPB committee had originally (1944) recommended the 170-mile and 85-mile distances for co-channel and adjacent channel interference with the understanding that the calculations neglected tropospheric interference. A study by the television association dated December 1944, which took into account theoretical predictions of tropospheric interference, recommended much greater separations, closer to 225 miles for co-channel interference. The competition between RCA and CBS over UHF and color added to the confusion about how to deal with this form of interference. CBS emphasized the seriousness of the problem in the VHF band; RCA generally played down the possibility by emphasizing the lack of experimental data. RCA and its allies were more interested in convincing the FCC that UHF was inferior to VHF because of the existence of other forms of interference in the higher frequencies—shadows and ghosts caused mainly by interference with buildings and other objects. CBS predicted that these problems would not be serious. The companies, with the support of their engineers, thus took advantage of uncertainty and a lack of definitive data to promote their corporate policies.[30]

The FCC made an important compromise in November to help realize the industry demand that New York City should receive seven stations. Specifically, it located stations "somewhat closer together in the eastern part of the United States," approximately 150 miles for co-channel stations and 75 miles for adjacent channels. The commission apparently assumed that since knowledge about interference was uncertain, the staff was justified in taking engineering risks in order to come up with a compromise plan satisfying the industry's desire for more stations in large cities and the commission's wish to distribute stations to small cities across the country. But the FCC did explicitly state in a December 1945 report that these distances were "subject to change as additional information concerning tropospheric wave propagation is obtained." Despite the uncertain factors that went into the development of the assignment plan, the commission assured the public in the same report that these standards were "based upon the best engineering data available, including evidence at hearings, conferences with radio engineers, and data supplied by manufacturers of radio equipment and by licensees of television broadcast stations."[31]

UHF Reconsidered, 1946–1947

By fall 1945, more than 115 applications for commercial television licenses were pending before the FCC. The process of assigning frequencies to stations was delayed while the commission considered a new request by CBS in September 1946 to authorize UHF broadcasting. This time, the company specifically wanted broadcasters to use the new band only with its color system, instead of with any other black-and-white system, including the high-definition black-and-white system. CBS executives testified that "our work in color has convinced us that no black and white picture, regardless of the number of lines, can compete with a color picture transmitted on the CBS proposed standard." The FCC needed to address the question of whether to authorize commercial operations in UHF using color standards because CBS had been successful in stifling the development of VHF television by arguing that it would soon become obsolete. By August 1946, eighty applicants had withdrawn their requests for commercial licenses, partly because of uncertainty about the future of VHF television using the black-and-white standards. The FCC had helped fuel this uncertainty when it announced in April that it might reevaluate its assignment plan by authorizing UHF and color. A few companies—including Zenith, Westinghouse, and Bendix—supported CBS by announcing that they would begin manufacturing television equipment using CBS color. CBS initiated a series of public demonstrations of its color system, beginning in January. The next chapter discusses CBS's attempts to gain authorization for its color system in more detail; the analysis here focuses on how the participants in the debate evaluated the UHF band, which held the potential to play a major role in helping the FCC set up a nationwide system of television stations.[32]

The commission denied CBS's petition for commercial use of UHF color in March 1947. This decision was mainly based on the FCC's belief that CBS's color system was not ready and that other rival systems under development might be superior. But the commission also argued that engineers needed to conduct further tests to determine the propagation properties of television transmissions in the UHF band. Regulators were concerned about potential interference problems in the UHF that might reduce transmission coverage, and they were convinced the industry needed to perfect high-power tubes and other equipment.

Although the FCC had conducted some of its own studies to learn more about broadcasting in the UHF, its final judgment to a large degree depended on industry testimony. Even in the FCC's own studies, experts did not work independently but collaborated with industry engineers, mainly because the FCC did not have access to the necessary equipment.

During 1946, the commission requested committees of the RTPB and the Radio Manufacturers Association to evaluate UHF color. A committee of the manufacturers association known as the Television Systems Committee had been established during spring 1945 to replace panel six of the planning board when it appeared that the nonprofit groups sponsoring the organization might decline further financial support. This fear proved groundless, however, and the association transferred the activities of the committee back to panel six in June 1946. Although the committees submitted complete reports evaluating the color systems, they could not make recommendations about UHF because of an apparent lack of data. The committee of the manufacturers association evaluating UHF reported after its meeting in January that it "did not have sufficient technical data or field-test experience to define how completely a service area can be covered under these conditions, nor did they know of the existence of such data." The planning board chairman reported in December that a determination of the adequacy of the UHF channels had been "on our agenda," but "no one saw fit to actually produce any information. . . . From a propagation standpoint, we did nothing." But experts had conducted a number of propagation studies as early as the spring of 1946, including a project coordinated by the FCC using CBS and RCA equipment to determine field intensities and regional coverage of television stations operating in New York City on a frequency of 700 MHz.[33] Why the planning board did not consider the results of these studies is unclear, but it appears that the members believed it was more important to commit their resources and time to evaluating the different color systems. The failure of the technical advisory groups to fully address the UHF problem played an important role in the FCC's final determination that experts needed to conduct more studies.

The commission also based its final decision on the testimony of RCA and DuMont during hearings in early 1947 that their tests demonstrated UHF might have serious problems. The FCC's report contended that these field tests "were sufficient to cause grave concern as to whether or not the ultra high frequencies now set aside for experimental television are really suitable for that purpose." The most serious problem indicated by the tests was that hills and buildings would create shadows that would interfere with UHF transmissions. Significantly, the critics of UHF did not deny that decent service would be possible using these frequencies; they mainly wanted to point out that the performance would not be as good as the service provided in the VHF band. DuMont engineer Thomas Goldsmith testified that UHF "can give a substantial broadcast service if properly engineered, but I don't believe it will ever reach the degree of coverage that is now available with the lower frequencies." Although CBS's petition had not requested that UHF replace VHF television, the companies with a major interest in the lower band feared this might happen if the FCC authorized UHF color for commercial service. Critics also successfully convinced the commission that experts needed to complete further tests to determine if frequencies higher than the 480 to 920 MHz band might be better for color television. A handwritten memorandum by an FCC staff member reported that there was a "school of thought which gained momentum during the hearing that the 480–920 mc band is too low for the eventual system." According to the author of the document, "these proponents believe that the disadvantages of both high and low frequencies are present at 700 MHz. . . . It appears then that an extensive propagation survey is needed before committing these bands permanently to television."[34]

The fundamental problem of having to formulate policy based on uncertain technical knowledge played an important role in the debate about the commercialization of UHF in 1947. CBS and the company's allies not only argued that their tests demonstrated the adequacy of the UHF band for television broadcasting but also pointed out inconsistencies in how some government and industry officials dealt with the problem of making policy based on limited knowledge. They contended that experts had completed "tremendous [*sic*] more field testing" for UHF broadcasting than had been done when the FCC approved commercial use of the twelve black-and-white television channels in the VHF band. Although Jolliffe and other engineers believed that this was a matter of "opinion," Commissioner Jett provided powerful support for CBS's position by asserting that it was "a known fact we did not have any information with regard to television broadcasting service from 150 to 300 megacycles at the time the service was made commercial [in 1941]."

Since Jett had served as the chief engineer on the commission when this earlier decision was made, his opinion could not easily be dismissed.[35]

Industry engineers were not always more consistent than the FCC in dealing with the problem of technical uncertainty. Although with certain earlier policy decisions, Jolliffe had called on the FCC to assume risks and not wait for exhaustive study, with this new decision he took a decidedly conservative stance, arguing that "knowledge of radio propagation in the 480–920 mc band and above is too meager to judge the characteristics and qualities of transmission in the upper frequency regions, with the degree of safety that is desirable for allocation of a permanent commercial service." In agreeing with Jolliffe, the final FCC report attempted to respond to the charge that it was using a different set of standards to judge UHF than it had used in 1941 to evaluate VHF. The report pointed out that "before standards were adopted for monochrome television, there were at least seven stations in operation in several cities and several thousand television receivers were outstanding." By contrast, according to the commission, all experimentation with UHF had involved transmissions from one location, New York City, and very few receivers had been available for testing.[36]

The testimony of manufacturers that engineers still needed to develop equipment using UHF was also an important factor in convincing the FCC to deny CBS's petition for commercial status of UHF. Representatives of Westinghouse and Federal Telephone and Radio told Jett during the 1947 hearings that they would not be able to deliver transmitting apparatus capable of producing the required power levels for television broadcasting throughout the ultrahigh frequencies until at least the following year.[37] In comparison especially with RCA, CBS was at a disadvantage because it was not a diversified company involved in both manufacturing and broadcasting. It had to rely on other companies to manufacture the equipment for its system.

Prefreeze Events

The FCC's decision not to authorize UHF color television helped stimulate interest in standard VHF broadcasting. The number of new applications for VHF channels increased dramatically beginning in fall 1947. In June that year, 11 stations were broadcasting regularly, another 65 had commercial licenses, and 9 applications were pending before the FCC. By September 1948, the number of stations on the air had increased to 36 (in twenty-one cities), an additional 116 had received commercial authorizations, and the commission was considering 304 new applications. While manufacturers had been able to sell only 14,000 television sets in 1947, during 1948 they sold 172,000.[38]

But the expansion of VHF television during 1947 and 1948 helped highlight technical and socioeconomic problems that needed to be resolved before the FCC could implement a satisfactory nationwide assignment plan. As early as August 1947, the commission was convinced by new industry and government studies to reconsider the proposal to allow fixed and mobile services to share operations on twelve of the television channels.[39] By 1948, all parties agreed that sharing would result in intolerable interference. As a compromise solution, the FCC gave television exclusive control of twelve of the thirteen VHF television channels; it placed fixed and mobile services on channel 1. Television stations already broadcasting on channel 1 were reassigned to one of the other VHF channels. Although television broadcasters were not pleased about losing one of their channels, they believed that twelve exclusive channels would be preferable to thirteen channels partially shared.

The fixed and mobile services used the reevaluation of the sharing plan as an opportunity to raise important questions about the criteria employed by the commission in allocating frequencies to different services. Their criticisms underscored how the FCC was tacitly defining the public interest in economic terms. Television was given preferential treatment because of its anticipated role in helping to create an important new industry. Representatives of the fixed and mobile services requesting more channels in the VHF band believed the FCC should give them priority in allocating frequencies because of their significant connection to public safety, a consideration that the FCC had traditionally valued first. "It requires a good deal of imagination to envision an occasion in which television could render services which are of an emergency nature," they pointed out: "it must be recognized that in its essence television is a luxury service, its principal use being in the field of entertainment." One witness at hearings in November 1947 tried to convince the commission to give two more of the television channels to mobile and fixed services, specifically for the development of mobile and point-to-point telephone service, which he believed would prove more useful than television in emergency situations.[40]

The mobile and fixed services felt they had been forced to try to acquire channels from television because the federal government refused to yield any of the VHF frequencies it controlled. Emergency services operated by state and local governments resented not being included in IRAC deliberations, which had authority over radio frequencies allocated to federal agencies. "All of us under the jurisdiction of the Federal Communications Commission," according to one witness testifying at a commission hearing in November 1947, "are having a most trying time to nourish and develop our bodies when the only food we have consists of the crumbs from the IRAC banquet table." The same person went so far as to call on all the nongovernmental services to "join forces, descend upon the Congress of the United States and demand an amendment to the Communications Act of 1934 which will broaden the control which the Federal Communications Commission will have over many of those frequencies now under the control of" the interdepartment committee.[41] But this proposal fell on deaf ears. The committee did not give up frequencies to the fixed and mobile services, nor did it include nonfederal emergency services in its deliberations. The television industry had also hoped it might be able to convince the government to give up VHF frequencies, especially in portions of the band between television channels 6 and 7. But the outbreak of the Korean War and renewed military demands for radio frequencies put to rest any lingering hope that the government might give up frequencies.

Although the suspension of sharing with fixed and mobile services eliminated one form of interference for television broadcasters, both cochannel and adjacent-channel interference between nearby television stations remained a problem. The loss of channel 1 actually made this problem worse; the FCC's television-station assignment plan of August 1947 decreased the geographical spacing between stations in order to fit the broadcasters using channel 1 into the twelve remaining channels. Increased pressure for more television stations in the largest cities and the need to accommodate new Canadian stations near the border with the United States led the FCC to further reduce geographical separations in a new plan presented in May 1948.[42]

During hearings evaluating this plan in June, witnesses presented new evidence indicating that the allocation plans had not sufficiently taken into account potential interference from tropospheric propagation. The FCC received numerous complaints about all types of interference, some of which was made worse because of unusually strong sunspot activity. The commissioners realized that they had made too many compromises in attempting to develop a nationwide system of television stations. Commissioner Albert Wayne Coy, who in January 1948 had replaced Denny as chairman, admitted that the FCC had "continually thrown away the 'safety factor' of greater mileage separations in a series of progressive steps." The commissioners decided that experts needed thoroughly to study all the problems involved, including especially the future of UHF television. Coy asked whether "we want adequate planning reflected in the television service or whether we are going to yield to the insistent pressures of applicants who are now willing to take whatever they can get." By announcing on September 30, 1948, a "freeze" on new applications pending a thorough study of all the problems involved in establishing a nationwide system, the FCC made it clear that it wanted to pursue "adequate planning." Although Coy thought six to nine months would be a sufficient period of time to reevaluate television policy, the freeze actually lasted nearly four years.[43]

Technical Advice and the Television Freeze, 1948–1949

After the September hearings in 1948, the commissioners conducted a number of engineering conferences to gather information from industry and government sources about television broadcasting. They were mainly interested in obtaining engineering data about propagation coverage and interference, but these discussions only underscored the complex relationship between technical issues and decision making on policy.

A new engineering advisory group that combined the institutional resources of the Institute of Radio Engineers and the Radio Manufacturers Association, the Joint Technical Advisory Committee, was established during 1948 to assist the commission with policy decisions. Chairman Coy first suggested replacing the RTPB with a new committee during a speech at the national convention of the Institute of Radio Engineers on March 23, 1948. Coy was critical of the planning board because he believed the different panels were each more interested in promoting a particular special service rather than the overall public interest. The new group helped to consolidate the work of preexisting committees of the Institute of Radio Engineers and the Radio Manufacturers Association. The joint committee made final decisions after evaluating all available information from different technical groups, including especially the preexisting allocation and standards committees. The boards of directors of the manufacturers association and the planning board formally established the joint committee on June 20; the RTPB was dissolved on July 1.[44]

Compared with earlier technical advisory groups, such as the system committee and the RTPB, the joint committee was more cautious and self-critical in its deliberations and in its interaction with the commission. The difference is especially striking when the role of the joint committee in the making of policy for television is compared with the strongly technocratic work of the engineers who helped establish a nationwide system of AM radio stations during the 1920s. Whereas the AM technical experts generally had not considered how technical decisions might be closely related to particular socioeconomic developments, the charter of the joint committee specifically "recognized that the advice given may involve integrated professional judgments on many interrelated factors, including economic forces and public policy." The charter still affirmed the important role of engineers, not simply because they were in the best position to evaluate engineering factors but because their professional training enabled them to "maintain an objective point of view" in considering all kinds of information. The Institute of Radio Engineers and the Radio Manufacturers Association each appointed four engineers from government and industry to serve as members of the joint committee. They based their appointments on the engineers' "professional standing, integrity, and competence." Philip Siling, who had served on the engineering staff with both the interdepartment committee and the FCC, as well as RCA, headed the committee; Donald G. Fink was vice-chairman. Most members, including David Smith of Philco and Jett of the FCC, had been involved in policy making on telecommunications for many years.[45]

One of the main goals of the joint committee was to "remove commercial bias" from information relating to television policy. Unlike the planning board and the system committee, the joint committee specifically "omitted from the committee records" the "business affiliations" of its members. Individual engineers also testified during various engineering hearings held by the FCC, but the FCC placed more authority on the reports of advisory groups such as the joint committee because of the perception that they could make judgments free of commercial bias. In a discussion of the testimony of engineers representing various companies, one staff member complained that "whenever we have an allocation hearing before the commission, . . . the people who come in and testify aren't experts any more, but they have been retained by particular stations or particular applicants, and you simply cannot get an expert judgment."[46]

Because it believed its members were of the highest professional and moral standing, the joint committee did not consider whether its work might be influenced by particular corporate affiliations. The commission did not question this action, despite the fact that DuMont, which had no representatives on the joint committee, publicly disagreed with some of the committee's conclusions—notably, the evaluation that experts needed to pursue further technical development before the commission authorized UHF for commercial broadcasting.[47] The commission's stand in this case differed from the position Fly had taken as chairman. Fly had established the system committee because he believed that the standards committee of the manufacturers association did not fairly represent all the major television manufacturers and broadcasters.

Unlike the earlier advisory groups, the joint committee tried to develop formal procedures that would take into account the problem of having to make policy decisions based on limited and uncertain knowledge. "To guard against misinterpretation of its data," the committee established three classes of technical information, "differing in the degree of reliance which can be placed upon them." Class A data was the most credible: it included facts or observations that most experts agreed were adequate and reliable. Class B information represented "engineering estimates . . . based on limited experience, or statements based on theory not fully confirmed." Class C data included speculation and conjectures "based on more or less arbitrary extrapolation from limited experience." The committee believed allocation and related policy decisions ideally should be based on class A data, but as a practical matter experts often had to use class B information.[48]

In response to an FCC request, the first study undertaken by the joint committee analyzed the feasibility of using UHF for commercial television. Donald Fink met with the commission in June 1948 and received a list of questions, including inquiries about the present state of development of UHF equipment, the amount of UHF-TV experimentation the industry had completed, the service areas and the amount of interference expected in these frequencies, how the government might assign UHF channels to stations across the country, and the costs of UHF-TV equipment. Fink sent the questions to various technical groups, most importantly the Television Systems Committee and Radio Wave Propagation Committee of the Institute of Radio Engineers, but also the Television Systems Committee and Committee on Television Transmitters and Receivers of the Radio Manufacturers Association. The joint committee evaluated the various committee reports it received and presented the FCC with its own report on September 20. The general recommendation that experts needed to conduct more observations and research before officials authorized UHF for commercial television supported the FCC's decision at the end of the month to freeze television licensing to allow a thorough review of television policy. Specifically, the joint committee reported that proper equipment for UHF television broadcasting would not be ready for at least one to three years. The committee also judged that experts could not formulate an assignment plan for UHF stations because technical factors such as the propagation properties of these frequencies were all in the class C category. They recommended that whatever the final decision about using the UHF frequencies, the commission should retain the twelve VHF channels to form the "backbone of the monochrome television system."[49]

In December, the joint committee presented the FCC with a report on propagation and equipment characteristics for VHF television. As with the UHF report, the VHF study was cautious and provisional. The committee emphasized that its computations of coverage and interference areas for VHF television stations should only be considered "illustrative and . . . conditional on more and more conclusions on propagation." The report admitted that "the evidence presented, while collected from as many sources as possible in the time available, does not necessarily represent all the data available in the industry." Partly because the committee acknowledged the complex relationship between technical and nontechnical considerations, the members wanted to avoid usurping the decision-making role of the FCC. They emphasized that the commission had to provide engineers with certain policy assumptions about allocation philosophy before they could recommend an ideal assignment plan. In testimony at an FCC engineering conference on December 3, Fink stressed that the joint committee "is not supposed to be . . . a partisan in this, and if anything I have said makes it appear that we are taking a stand for one theory of allocation as against another, I wish to correct that impression completely." An important issue that the joint committee wanted the commission to address was the amount of interference that should be considered acceptable. The committee emphasized that this decision partly depended on how the commission evaluated subjective factors such as the "psychology of the typical" viewer. To illustrate this point, the committee referred to a viewer interested in watching a televised boxing match featuring Joe Louis (the committee's meaning is clear if not its syntax): "If the only time he ever wants to see Joe Louis fight it happens that everything is OK. But if his service depends upon a distant station and it is not on, then his psychology is different." To help gain a better understanding of the subjective aspects of interference, the joint committee sponsored a series of tests at RCA Laboratories during 1949 using one hundred observers as test subjects.[50]

The qualified and prudent nature of the testimony of the joint committee underscored the fact that officials had to address important compromises involving a combination of factors before the government could establish a nationwide assignment system dictating the location and frequencies of different stations. The FCC had to consider a number of important issues, including how to balance the need for exhaustive tests and observations with the demand for timely decisions to stimulate development; whether to use both VHF and UHF frequencies and, if the decision was to do so, whether to assign both kinds of stations to the same city; whether to accept compromises well short of total freedom from interference; whether to provide television service to all of the country or only to major urban areas; whether to authorize a small number of high-power stations providing large service areas or establish more low-power stations providing a variety of programming.[51]

Partly in response to the joint committee's acknowledgment that its reports were not based on all available information, during 1949 government engineers took a more active and independent role gathering and evaluating data necessary for television policy. This development reflected the growing importance of field studies conducted by radio engineers at the Bureau of Standards and the FCC after the war. During the engineering hearings at the end of 1948, the commission appointed an "Ad Hoc Committee" with a mandate to evaluate "propagation problems left unsolved at the engineering conference[s]." In the beginning, the membership included two engineers from the Bureau of Standards, three from the FCC, and four from independent consulting firms; in the spring of 1949, however, six engineers from industry were added (two from CBS, two from DuMont, one from RCA, and one from Westinghouse). In comparison with the joint committee, the ad hoc committee was more representative of the entire television industry. With the FCC's chief of the Technical Information Division as chairman, the ad hoc committee held twenty-eight formal meetings from November 1948 through July 1949. The Information Division had a mandate to help the commission's Bureau of Engineering gather data about radio propagation and related subjects. In support of the ad hoc committee, the Information

Division cooperated with the General Radio Propagation Laboratory of the Bureau of Standards to complete studies of field intensity and radiowave propagation. According to the FCC, the resulting reports that integrated these government studies with industry findings probably represented "the most highly scientific study which the Bureau of Engineering has ever undertaken." The reports covered such subjects as the effects of terrain and other surface features on wave propagation, the effects of different antenna heights and transmitter power on television coverage, and the amount of interference caused by tropospheric and ionospheric propagation at different frequencies.[52]

Despite the large amount of work undertaken by the so-called ad hoc committee, the major allocation report in June 1949 echoed the tentative conclusions of the joint committee. Members of the new committee admitted that they still did not have a thorough understanding of VHF and UHF propagation under all conditions. Their report warned that officials should not rely on the "voluminous treatment of the meager data" to "lend an air of authority to what is at best an interim solution."[53] Individual committee members expressed even more skepticism about the final recommendations. Thomas Carroll of the National Bureau of Standards argued that the committee's evaluations of such technical issues as tropospheric interference and rough terrain propagation were still only at the "engineering guess level" (class C data, according to terminology of the joint commission). He did not want the FCC to establish rigid rules and standards of engineering practice based on these "guesses" because that "might throttle the possible development of TV broadcasting in places where our guesses turn out to be wrong." But Carroll and other committee members also argued that the lack of data should not be used as a justification for continuing the freeze and keeping stations off the air. These engineers wanted the commissioners to adopt a "flexible" and "conservative" assignment plan for television stations that they could adjust once stations were on the air and experts could collect actual propagation data. They believed it would be much easier to modify at a later date an assignment system that overestimated the amount of separation needed between television stations than one that proved to be too compact. According to Carroll, "confessions of ignorance right now" would be "more likely to lead to a flexible policy which can gradually be brought into accord with the laws of nature propagationwise, which laws we unfortunately have to discover first."[54]

The FCC proposed a revised assignment plan in July 1949. It incorporated the ad hoc committee's recommendation that stations should be spaced far apart. The proposal called for conservative co-channel station separations averaging more than 200 miles and ranging up to 328 miles. The commission proposal authorized forty-two UHF channels for television broadcasting using standards identical to the VHF. Any new color-television system would need to have to fit into the established 6 MHz channels. The commission also emphasized that it did not intend to move VHF stations to the UHF band. Future stations would operate on both VHF and UHF, but the July 1949 plan allowed most of the VHF stations already authorized to continue using the same frequencies. Finally, the commission recommended intermixing UHF and VHF stations in the same city in order to stimulate interest in dual-band UHF/VHF receivers.

The commission stressed that its proposed assignment plan was based on sections 1 and 307(b) of the 1934 Communications Act, which in turn were based partly on revisions of the Davis Amendment of the 1927 Radio Act. The 1934 statutes required the FCC to endeavor to provide broadcasting to all citizens of the United States and to distribute facilities on an equitable basis to different sections of the country. Following these guidelines, the commission used four priorities to establish an assignment table. The first priority emphasized that all parts of the country should receive at least one television station. Like government officials during the 1920s who helped establish the assignment plan for AM broadcasting, the commission emphasized the importance of providing rural service. The second goal sought to provide at least one station for each community. For the third priority, the commission stressed that it would try to provide competition by authorizing at least two services to all parts of the country. The fourth priority was to give each community a choice between at least two television stations.[55]

Whereas the 1920s AM assignment plan had been based on technocratic assumptions that helped insulate its establishment from political controversy, the effort to develop an assignment plan for television during the freeze was less technocratic and much more controversial. The admission by engineers advising the FCC that they could not give authoritative advice helped open the freeze to contentious debate. Key companies, especially DuMont Television, objected to the July 1949 plan and subsequent commission proposals. More important, members of Congress, especially Senator Edwin Johnson, Democratic chairman of the Senate Committee on Interstate and Foreign Commerce, also criticized FCC actions and helped prolong the freeze until 1952. The political pressure on the commissioners led them to seek more technical data about tropospheric interference and other factors before lifting the freeze, instead of making final decisions—as they had often done in the past—based on provisional, but seemingly authoritative, recommendations.

Senator Johnson's major concern was that the public should have access to all technical advances in television, including, especially, color television. By insisting that the FCC include the color issue with the assignment problem, he helped extend the freeze for at least eighteen months. The next chapter will examine in more detail how Johnson encouraged the FCC to investigate the patent structure of the companies developing color television as well as company actions that might lead to monopolistic control. The old issue of "RCA's alleged undue influence in commission's policies" was revisited during the public hearings dealing with color television.[56]

Johnson was also concerned about the FCC's actions regarding UHF, particularly the decision by the commission to retain the VHF band, instead of placing all broadcasters on an equal level in the UHF. "I regard it as tragic for the ultimate development of television," the senator lamented in November 1949, "that the VHF allocations heretofore made is handicapping the adoption of a truly equitable and scientifically practical VHF-UHF allocation." During the allocation hearings in 1945, the FCC had promised to move television to the higher frequencies. After signaling for years that it might not follow through with this promise, the commission made its intention clear in March 1949 when chairman Coy addressed a meeting of the Advertising Club of Baltimore.[57] With the decision to retain 6 MHz channels, the original idea of using wide channels in the UHF to introduce a higher definition system was also abandoned. Color television, which originally seemed to demand wider channels, was still possible since, by 1949, both CBS and RCA had found ways to fit their color systems into 6 MHz channels.

Although the industry conducted some research during the freeze to prepare UHF for commercial use, most manufacturers were more interested in selling VHF television receivers to consumers anxious to watch broadcasts from the prefreeze VHF stations that continued operations during this period. One notable exception was Zenith, which made preparations soon after the beginning of the freeze to manufacture and sell dual-band UHF/VHF receivers. Other manufacturers harshly criticized Zenith when it launched an advertising campaign encouraging consumers to buy dual-band receivers in order to ensure protection against possible equipment obsolescence after the commission lifted the freeze and authorized UHF broadcasts. The FCC did not strongly push manufacturers to prepare for UHF broadcasting. However, it did encourage efforts of some manufacturers, including RCA and Philco, to build and operate, during the freeze, experimental UHF stations. At least one commissioner, Frieda Hennock, the first woman appointed to any of the federal regulatory agencies, also seemed to support Zenith's position by stressing the need to protect consumers from obsolescence. Hennock called for new legislation clearly authorizing the FCC, in consultation with the Federal Trade Commission, to take action warning the public about "the uncertainties inherent in the purchase of any particular television receiver." She suggested that Congress give the FCC authority to warn consumers about potential decisions relating to standards or frequency allocation that might affect the value and usefulness of television and radio equipment.[58]

Allocation Proposals, 1949–1950

Not surprisingly, given the competitive nature of the industry, the response of individual broadcasters and manufacturers to the FCC's assignment proposal of July 1949 depended partly on whether or not they might benefit from any change. The established broadcasters and networks—mainly NBC, CBS, and the prefreeze VHF stations—generally supported policies that would not disrupt their dominant position. Although they did not like some aspects of the plan, they mainly refrained from criticizing it because the proposal made few changes that would adversely affect their status. By contrast, new applicants and the two smaller networks, ABC and DuMont, promoted policies that would help them break into new markets.[59]

Allen DuMont, a manufacturer and station owner as well as network organizer who had close ties to the independent and opinionated Edwin Armstrong, was the strongest opponent of the FCC's proposal and presented the most complete alternative plans. Although the government assignment would allow the prefreeze VHF stations to continue using the same frequencies, DuMont favored forcing a number of VHF stations to switch to the new UHF band. DuMont and his allies also opposed the policy of intermixture; they were convinced that UHF stations would never be able to compete with established VHF stations operating in the same market. Instead of trying to maximize the amount of area covered by different stations, the DuMont plan placed a higher priority on maximizing the number of people who could receive broadcasts. Using this criteria, DuMont, in his proposal, was able to authorize more VHF stations (approximately four) in all the major cities.

Many of the established VHF stations in the large cities had network affiliations with NBC or CBS. In spite of the freeze, 106 VHF stations were operating in sixty-four communities by the end of 1949. Fifty-nine of these stations had been given special temporary authorizations after the freeze was placed into effect. Altogether during this period, the FCC estimated that 57 percent of the population of the country had access to television broadcasts. The FCC's plan generally did not allow more of the prime VHF stations to operate in the largest markets. Any aspiring new network would need affiliation with these stations to be competitive, at least until UHF became better established. Both DuMont and ABC thus favored a policy that would authorize more VHF stations in each of the major cities. They argued that this would create more competition and give the public more programming choices.

Using elaborate nine-foot-by-sixteen-foot maps illustrating the differences between the FCC's proposal and his plan, DuMont tried to convince the commission to reconsider the issues. In February 1950, he presented an even more elaborate and complete nationwide plan that made station assignments in the top 1,400 markets in the country.[60] Unlike the case of the AM-assignment decisions during the 1920s, a number of the participants debating the establishment of a television-assignment plan, especially opponents of the FCC's plan, were much more willing to acknowledge and explore the socioeconomic implications of technical decisions. Rather than emphasize only how their plans were justified based on technical demands, these participants also stressed the economic, political, social, and legal grounds for their views.

From October 1950 through January 1951, the FCC conducted crucial hearings on all aspects of television assignment, including the value of the alternative proposals presented by DuMont and other broadcasters. The ad hoc committee and the FCC's Technical Research Division also introduced important results from further studies they had helped sponsor and organize. The Technical Information Division was renamed the Technical Research Division in October 1949; the chief continued to head the ad hoc committee and members of the division spent a "considerable amount of time" preparing the new reports of the ad hoc committee that were presented during the commission's hearings in October 1950. Engineers collected new propagation data from monitor-

Staffers work on one of the large display maps DuMont used in 1948 in his unsuccessful attempt to convince the Federal Communications Commission to adopt his television assignment plan.

ing stations with the assistance of the Central Radio Propagation Laboratory of the National Bureau of Standards. Collaboration made sense since different departments of the government, including the military services that used data from the central laboratory, shared a common interest in gaining a better understanding of UHF and VHF radio propagation. Staff members at the Bureau of Standards were careful to emphasize that they were not advocating "any policy questions in this field." They drew a sharp distinction between their technical and scientific work and "policy questions" that were "clearly the function of the Federal Communications Commission." One representative testified that "our study was intended to be an objective analysis leading to quantitative criteria measuring the degree to which various allocations achieve the goal which the commission proposed to be of highest priority." The two government agencies also collaborated with industry to obtain needed data. When NBC and RCA conducted extensive measurements of broadcasts from the first station to transmit regular television programs on UHF—located in Bridgeport, Connecticut—the FCC's Technical Research Division helped plan the project and allowed the participants to use its monitoring stations. The division also donated and installed equipment at the University of Connecticut that was used in support of the project by members of the electrical-engineering department.[61]

The FCC needed new data about UHF and VHF to help evaluate different assignment proposals. Discussion of the DuMont plan during the commission's hearings in the fall of 1950 underscored the complexities and controversies involved in developing a nationwide assignment plan. The commission believed that its plan was justified based on the list of priorities it had established that were demanded by key statutes in the 1934 Communications Act, including the need to provide equal service to all sections of the country. But DuMont contended that other interpretations of the law were possible; in particular, his interpretation would more fully uphold the intent of the key statutes—to make sure television would be responsive to democratic values and benefit all citizens, not simply a select few.

DuMont contended that his plan was more democratic since it was based on qualitative rather than quantitative considerations and valued people over territory. The first "more specific" priority that he believed would meet the statutory objectives was "to remove every vestige of monopoly or oligopoly in the broadcasting network industry." Other priorities dealt with the need to avoid intermixture and the importance of providing at least four channels to as many metropolitan areas as possible. DuMont also argued that even when evaluated according to the FCC's quantitative priorities, his plan was superior to the commission's proposal. In detailed testimony, DuMont representatives pointed out that their plan would better satisfy the FCC's first priority by providing television broadcasts to an additional two hundred thousand square miles of territory. And because it would provide service to forty-three more communities than the commission's plan, the DuMont proposal seemed to do a better job fulfilling the second priority of assigning a station to every community in the country.[62]

The superiority of DuMont's plan partly resulted from the fact that it used more UHF channels than the commission's proposal—sixty-nine as opposed to forty-two. DuMont's finessing of the commission's priorities mainly demonstrated the limitations of quantitative, technocratic thinking for providing definitive analysis; his actions also showed that the FCC, in developing an allocation and assignment philosophy, needed to take into account essential qualitative considerations, such as intermixture and potential monopoly control.

Archival records indicate that the FCC did consider some of these qualitative issues when it developed its own assignment plan and rejected DuMont's alternative. The engineering department, in particular, played a key role providing expert advice. Although the engineers on the ad hoc committee and the Joint Technical Advisory Committee had avoided usurping the decision-making role of the commission, arguing that "in the absence of specific knowledge, certain assumptions must be made by someone if the commission is to proceed with the allocation of television stations," one influential member—the chief of the FCC's Technical Research Division—also emphasized that, in his judgment, it was "preferable to have these assumptions made by persons fitted by training and experience to make them, rather than by popular vote or by argument and compromise." Intermixture was an important policy assumption that the FCC's engineers strongly supported. Although they acknowledged DuMont's concern that "economic problems will be faced by new UHF broadcasters in the VHF TV areas," they believed this would be only a short-term problem. The staff favored intermixture as a way to stimulate manufacture of dual-band UHF/VHF receivers, and they argued that without intermixture "it would be necessary to limit many areas to one or two VHF stations."[63]

Intermixture was a controversial issue; a number of broadcasters and manufacturers testified in support of this policy at the commission's hearings late in 1950. The position of the networks did not necessarily seem to correspond to their economic position in the industry. ABC, the third-ranked network, did not support DuMont on this issue; CBS did, NBC did not. The ad hoc committee and the joint committee avoided taking a stand either for or against intermixture; they argued that neutrality was necessary "because there are all kinds of economic nontechnical factors involved, competitive reasons, particularly." The commission claimed that "the majority of witnesses" opposed DuMont on this issue. A memorandum from the engineering department also indicated that the staff generally favored the status quo. According to the memorandum, the plan "was developed in a manner so as to cause a minimum of disruption to existing licenses." Without an activist like James Fly as chairman, the FCC's engineers and staff were unwilling to attempt major change. As we have seen, by rejecting intermixture, DuMont's plan specified the conversion of a number of well-established VHF stations to the UHF band.[64]

Despite not explicitly including the monopoly issue in its list of priorities, the commission did take this qualitative, socioeconomic concern into account when it evaluated assignment proposals involving the use of "stratovision." Stratovision, developed by the military during the war, ideally would use high-altitude airborne transmitters to broadcast television to all sections of the country, including rural areas in the West that would otherwise be outside the range of land stations. Westinghouse conducted experiments throughout the late 1940s and early 1950s to test its feasibility. In 1948, the FCC denied a request from Westinghouse to use stratovision with a VHF channel in Pittsburgh. The use of stratovision with UHF channels seemed more promising. Theoretically, it would provide nationwide television coverage and "provide an impetus to the growth and development of UHF" by taking "UHF out of the Cinderella class and placing it in a position of general respect next to VHF." The proposed use of stratovision for television broadcasting reminded participants in the policy debates about proposals for clear-channel and high-power broadcasting for AM radio during the 1920s. Just as a handful of observers opposed these techniques because of their potential for concentrating power in a few hands, a number of small broadcasters and influential members of Congress, including Senator Johnson, worried about the "potential monopolistic features of Stratovision." The FCC assured Senator Johnson that in evaluating stratovision it would not only take into account "the technical problems" but also "the economic and social problems which are implicit in the system." The strong opposition against the new method led the FCC to decide against supporting stratovision. The engineers concluded based partly on "economic or political considerations" that "the [assignment] plan must rely upon ground based stations."[65]

Another important "nontechnical" consideration that the commission took into account in its deliberations was educational television. The development of AM and FM radio had demonstrated that unless the FCC gave educational broadcasters special treatment, they would not be able to compete with the "better heeled commercial interests." Supporters of noncommercial educational television—including the Joint Committee on Educational Television, an organization representing seven educational groups, and the U.S. Office of Education—lobbied the FCC for this special treatment.[66]

During the hearings in November 1950, the educators requested that the FCC set aside one VHF channel for noncommercial educational television in each metropolitan area and in each major educational center. Since the FCC had denied a similar request in 1945, the representative of the Office of Education used the hearings as an opportunity to remind the commission of earlier precedents for the support of education by the federal government, including the establishment of the land-grant universities, many of which were now interested in establishing educational television stations. Both the act of setting aside land and the act of setting aside channels, he argued, "rest on the same fundamental notion that the public interest is best served when the need of the people for universal access to good education guides governmental action." He pointed out that although education "is assured of access to the use of the printed word because there is no limit to the number of presses which may operate," the limited number of channels available for television means that direct government intervention is necessary.[67]

The FCC's decision to set aside a band of frequencies for educational FM radio had helped put an end to the decline in nonprofit educational radio stations. Before the FCC made this decision, only thirty such stations had existed in the country; by 1950, the number had increased to more than one hundred. The educators argued that television offered unique opportunities for audio-visual education. By 1950, one college-owned television station had already been established; a number of colleges had begun planning for television and more than fifty more had expressed an interest in building stations.[68]

With the crucial support of Commissioner Hennock, the educators succeeded in persuading the commission to give preferential treatment to educational television in the assignment plans developed after the fall of 1950. The commission did not set aside a specific block of frequencies for educators, as it had done for FM, but it did reserve specific VHF and UHF channels in major cities. Traditionally, educators had not been able to compete with commercial broadcasters who had superior economic and technical support. During the 1920s, policy makers had used technocratic arguments to justify the decline in nonprofit educational stations. By contrast, during the commission's hearings in the fall of 1950, Hennock specifically responded to other commissioners' criticisms of educators for not providing the FCC with technically sophisticated plans by pointing out that the educators represented the public interest, which the commission should be supporting with special technical assistance: "We are the engineers; we have got the staff; these educators have not. They have not the money to hire these engineers until they are ready to go and build that station." Not all commercial stations and networks looked favorably on the idea of setting aside channels for educational stations. DuMont, whose assignment plan sought to maximize the number of VHF channels available for new commercial networks in the major cities, criticized the action because it would serve to limit further the number of these channels available. By contrast, CBS and NBC, which already had well-established VHF stations in the largest metropolitan areas, supported the idea partly because it would help restrict competition.[69]

Freeze Lifted: New Assignment Plan Implemented

After nearly four months of testimony from different witnesses about alternative assignment proposals, the commission spent an additional two months studying the voluminous record before issuing another preliminary report on March 31, 1951. The FCC's "Third Notice of Further Proposed Rule Making" went beyond earlier reports or notices by proposing a more complete, nationwide assignment of television stations to VHF and UHF channels. Industry reaction to this plan was not entirely favorable. Interested parties filed more than fifteen hundred comments with the FCC in response to the new proposal. Most comments focused on narrow aspects of the plan. DuMont again presented the most complete alternative proposal. The main difference between DuMont's new plan and his old proposal resulted from an acknowledgment that the commission would not compromise on intermixture. Otherwise, the new DuMont plan still insisted that the number of channels should be based on population or economic support instead of geographical area. DuMont also continued to call for four channels in as many cities as possible, which meant that the FCC could not make any special provision for educational stations. While the FCC plan provided 557 VHF stations in 342 cities, DuMont's proposal made provision for 655 stations in 375 cities. To help calculate all the assignments and give authority to the plan, DuMont used an electronic computer at the Massachusetts Institute of Technology. During the spring and summer, he developed and exhibited to the press and government officials an electronic display map with lights indicating locations of assigned channels and other relevant information.[70]

It took the FCC until March 1952 to evaluate all assignment possibilities, untangle contentious legal issues, and finally agree on a plan for lifting the freeze. The "Sixth Report and Order," issued on the last day of the month, continued to insist on most of the assumptions already put forward in earlier proposals, including intermixture and a determination of station distributions based on geographical area. The report rejected the major elements of DuMont's plan. The commission did agree to reduce station separations—partly by authorizing the use of new technical developments, notably offset-carrier operation—but not to the degree requested by DuMont and other broadcasters. The "Sixth Report and Order" presented a nationwide assignment table for the twelve VHF and seventy newly authorized UHF channels (between 470 and 890 MHz). Whereas the old assignment table for VHF-only television gave 400 assignments in 140 metropolitan centers, the new UHF/VHF table made available 2,053 assignments for 1,291 communities. The assignment made provision for noncommercial educational stations in 242 cities. The commission forced only a limited number of the prefreeze stations to make major frequency changes. The plan used intermixture extensively, especially in the major cities. Of the top 162 markets, 123 were intermixed.[71]

The commission felt confident that the nationwide assignment plan would create conditions conducive to the growth of UHF. All the commissioners except Robert F. Jones reassured broadcasters in the "Sixth Report and Order" that "the UHF band will be fully utilized, and that UHF stations will eventually compete on a favorable basis with stations in the VHF." They believed that the technical differences between UHF and VHF services that might prove an impediment to fair competition would eventually be solved by advances in "American science." Although some observers worried that UHF television would end up like FM radio, the commissioners believed the two developments were completely different; they pointed out that FM, unlike UHF, had to compete with a "fully matured competing service," AM radio.[72]

In contrast to this optimistic view of the future development of television in the United States, particularly UHF broadcasting, Commissioner Jones's dissent from the report raised serious questions. Like DuMont, Jones—a small-town Ohio Republican suspicious of the activities of big business in broadcasting—did not think the commission's plan would provide the proper conditions for UHF broadcasters to compete with VHF stations. He also believed the FCC could have reduced station separations in order to allow more cities to have their own television stations.[73]

Especially important for this discussion is Jones's contention that the FCC's justifications for its decisions were based on inherently flawed technocratic assumptions. Although the commissioners argued in the "Sixth Report and Order" that "healthy economic competition" would result from the assignment plan, the report implied that this was the case because the plan was based on the "best engineering information available," not because the FCC had made exhaustive economic studies. According to Jones, the commission's report "gives the implicit impression that engineering has dictated this unique plan." On the contrary, "engineering considerations do not determine a unique allocation. Thousands of different plans could be drawn up which were correct engineering-wise. . . . The engineering only places limitations on what can be done." Jones was especially upset that the commission refused to consider narrower station separations by insisting that its decisions were based on engineering data, despite the fact that engineering studies did not support the commission's conclusions. Specifically, he believed safety factors that the commission added to VHF station separations were unnecessary: "it is apparent that the commission's 'safety factor' is simply an increase in mileage separations arbitrarily imposed without any propagation data to support it in the VHF." This action seemed irrational to Jones. The fact that the fixed and rigid assignment table would not allow any adjustment of separations that included the safety factors contradicted the entire reason for having safety factors, which were imposed with the understanding that separations would be changed once new data became available. He also criticized the other members of the commission for using a double standard by not applying a safety factor for UHF station separations, despite the fact that the UHF "propagation data by contrast is almost non-existent. . . . The commission provides a 'safety factor' where the information indicates it is not needed (in the VHF) and they don't provide it in the UHF band where the information is so meager it might be advisable."[74]

Jones thus criticized the other commissioners and the FCC's staff for not acknowledging the complex relationship between the technical aspects of the assignment plan and policy considerations, including not only safety factors but also intermixture and the importance of assigning stations based on geographical area rather than population. If they had done a better job exploring the implications of these assumptions, he believed, they might have more fully established proper and fair conditions for UHF to compete with VHF stations. For example, if the commission had sufficiently taken into account the economic implications of its decisions, it might have recognized that the different minimum station separations for UHF and VHF, 150 miles as opposed to 170 miles, would place UHF television at a severe economic disadvantage with respect to VHF stations. UHF stations would have to purchase much more expensive equipment than VHF stations to "cover substantially the same number of locations in the . . . service area."[75]

Developments after the FCC lifted the freeze in 1952 and established the new assignment plan proved the accuracy of some of Jones's criticisms and predictions. While the number of VHF stations nearly doubled between 1954 and the end of the decade, the number of UHF stations fell from more than 125 to fewer than 80. Intermixture did not create conditions conducive for UHF stations to compete with VHF broadcasters. VHF stations had superior economic resources, an established listening audience, and a larger broadcast range. Theoretically, the FCC authorized UHF to operate using 1,000 kw power, as opposed to 316 kw for VHF channels 7 to 13 and 100 kw for channels 2 to 6. But UHF transmitters able to produce maximum power were not available until the end of 1954: in 1955, only a few were operating, partly because most UHF owners lacked the necessary economic resources. Even at maximum power, UHF did not have the range of VHF stations, although the exact difference in performance was unclear at the time. This reduced coverage was significant for advertisers interested in gaining access to the widest possible markets.[76]

UHF receivers needed to be developed and sold quickly for stations to compete with VHF rivals, but after the FCC lifted the freeze, shortages of UHF equipment became a problem. Also, poor-quality receivers, especially compared with VHF sets, limited the range of UHF stations. The industry had equipment available for converting VHF receivers to pick up UHF transmissions, but generally at a high cost. When backlogs of UHF equipment developed, manufacturers blamed consumers for not taking an interest in UHF television; however, manufacturers made little effort to promote UHF receivers through extensive advertising and marketing campaigns, especially in intermixed regions.[77]

When the DuMont network failed in 1955, DuMont's earlier effort to convince the FCC to authorize four or more VHF stations in many cities seemed justified. The 1952 commission's plan established only seven communities with four VHF channels, too few to sustain a fourth network. ABC also had difficulty surviving, especially since most of the VHF stations were already affiliated with CBS and NBC. The networks generally did not offer affiliation to UHF stations. This inability to gain network affiliation and benefit from national advertising further hurt the position of UHF stations.[78]

During the first few years after the release of the "Sixth Report and

Order," commissioners spent much of their time evaluating applications and implementing the assignment plan. Pressure from broadcasters and manufacturers as well as politicians and consumers to allow for the expansion of television broadcasting, which had grown during the freeze, continued during the first few years after the FCC lifted the freeze. The commission retreated from activist policies and generally did not worry about the overall state of the industry. By the late 1950s, when the plight of UHF broadcasters became clearer, Congress tried to step in to pressure the FCC and the industry, especially the networks, to look for ways to reverse the decline of UHF television. But many stations had already gone under; since the government did not institute new policy, such as complete de-intermixture, this trend was not dramatically reversed.

Like the assignment plan for AM radio developed during the 1920s, the assignment plan for television developed during the 1940s and early 1950s also had an important influence on the character of the broadcasting industry. Both plans helped reinforce the network system. The AM plan provided the proper conditions for the growth of networks; the television plan helped limit the total number to fewer than four. In both cases, we see technocratic tendencies, including an emphasis on the important role of engineers and the belief that regulation should primarily focus on technical issues and rely on instrumental rationality. This perspective often provides participants with a powerful legitimation strategy, whether clearly justified or not.

But there were major differences between the two developments. Officials explicitly made decisions affecting television, including the allocation of the radio spectrum to different services, in order to avoid the lack of planning that had contributed to the early problems with AM radio. In authorizing an assignment plan for television stations, the commission was more willing to take into account important qualitative, socioeconomic considerations such as the value of educational broadcasting. Further, the outside advisory groups established to help the FCC develop policy for television were more interested in exploring the implications of basing policy decisions on uncertain and provisional knowledge.

The blurring of the meaning of *technical,* especially when defined with respect to other factors, was connected to this development. As a result, the advisory groups were more cautious about providing advice to the commission and tried to avoid usurping the commission's primary duties. At the same time, government engineers played an increasingly important role. This development reflected the growing support for studies of radio-wave propagation and related radio and television research organized and undertaken by the FCC and the Bureau of Standards, with industry cooperation. The reliance on government engineers might potentially solve the commission's dilemma of whether to trust the testimony of engineers working for companies with an economic interest in policies being considered. Since they were able to conduct some of their own research, government engineers were less dependent on, although by no means entirely independent from, industry advice. Government engineers also played an important role convincing the commission to support controversial decisions such as intermixture. They believed their special problem-solving abilities and commitment to objective, unbiased decision making placed them in a unique position to make these decisions.

Because of the recognition that the engineering evaluation necessary for decision making on policy was often based on uncertain knowledge, commissioners were less likely to take "engineering risks," especially in comparison with earlier periods. This development contributed to the delay in the authorization of UHF television. The FCC, including the engineering staff, was also less likely to make decisions that would hurt the economic position of entrenched interests in the broadcast industry. This was especially evident in the refusal to convert major prefreeze VHF stations to UHF and in the decision to support a policy of intermixture. The next chapter explores some of these same themes in the context of the decisions leading to the authorization by the FCC of color television.

6)))

Competition for Color-Television Standards

Formulating Policy for Technological Innovation, 1946–1960

I believe television will not be a full-grown industry until color is provided. Color excites one of our most responsive senses. A travelogue in color, an oil painting reproduced in color, an advertisement for colorful clothing in color—what a difference in enjoyment the TV viewer would get.

Commissioner Robert F. Jones to Senator Edwin Johnson, February 2, 1949

Controversy also marked decision making for color television. In 1950, after rejecting all of CBS's earlier requests to authorize its color system for commercial operations, the Federal Communications Commission (FCC) adopted the company's design for using field sequences (see chapters 3 and 5). But three years later the commission overturned this ruling and chose another system supported by RCA and most of the rest of the television industry. Critics of the 1950 color decision argued (as did the commission's critics in the case of FM radio, discussed in chapter 4) that the FCC had made a major engineering mistake.

Many observers portrayed color television as being, like FM radio, an inevitable step in the progress of science and technology. Supporters believed it was inherently superior to monochrome television; they expected television to follow the same path as photography and motion pictures, which had pioneered the use of color. CBS, having developed the first potentially marketable system, led the way in promoting the benefits of the new technology. The company contended that the introduction of color technology would result in higher-quality programming and increased public interest. According to a CBS news release from 1940, in addition to providing "more pleasing lifelike" pictures and enhancing the "dramatic quality" of broadcasts, color would actually increase "the apparent definition of the picture and make small objects easier to recognize." It would add "depth to the whole picture" and eliminate "the flat quality that many people have felt exists in black and white television." CBS also contended that advertisers would be able to sell more products using color television and that "the cultural and educational scope will be increased."[1]

Other companies with strong economic investments in monochrome television played down the benefits of color, however. RCA engineer Alfred Goldsmith argued that the addition of color was "neither essential nor pressing." Both RCA and the Farnsworth company used the example of motion pictures to support their position. Farnsworth believed "there is ample evidence in the motion picture industry that monochrome pictures give entirely satisfying programs." "Even though color motion pictures have been available for many years," the company contended in 1946, "they still form a small percentage of the total motion picture releases." RCA and Farnsworth as well as most other manufacturers were anxious to profit from the industry's prewar commitment to black-and-white television. Farnsworth feared that "the use of color television . . . would handicap the earliest possible realization of an economically sound system" based on the available monochrome technology.[2]

These comments underscore how closely the development of color television was tied to business strategies; these strategies in turn had to be evaluated by the FCC to determine if they were consistent with the public interest. CBS's promotion of color needs to be understood in the context of its attempts to gain an economic advantage over RCA and other companies that controlled the patents to monochrome television. The company sought to use technological innovation to acquire a competitive patent position for a set of standards authorized by the FCC. This chapter thus analyzes how the commission evaluated the different proposals for color television. The commissioners felt obligated to take an active role in the development of color because of the 1934 Communications Act, which directed them to take actions encouraging technological innovation.

Many of the same issues analyzed in earlier sections of this book also need to be explored here, including the relationship between standardization and commercialization, the tension between advocacy and objectivity, and the role of technical experts and technical evaluation in policy making. We will be particularly interested in comparing policy making for color broadcasting in the United States through 1950 with the establishment of system standards for monochrome television before World War II and in analyzing the FCC's efforts to predict technological development and use these predictions to guide policy for color television.

CBS and Color Television, 1940–1946

Industry and government evaluation of CBS's color-TV system before the war provides an important background for understanding postwar decision making. Although the FCC did not authorize CBS's field-sequence system in 1941, both the commission and the National Television System Committee (NTSC) were impressed by CBS's presentations; the FCC approved the committee's recommendation that color television should "be given a six-month field test before standardization and commercialization." Members of the system committee who witnessed CBS's demonstration of its color system compared it favorably with black-and-white television. Thirty of thirty-four members claimed they preferred CBS's color to monochrome. Thirty-two thought that the quality of CBS's color was sufficiently high to present to the public. Thirty-three also believed the brightness of the color demonstration was acceptable. Only three members argued that all receivers should be compatible to receive both color and black-and-white broadcasts.[3]

Despite the favorable evaluations of the CBS presentation, the committee recommended that the FCC should not authorize CBS's system for commercial broadcasting until the company proved it could broadcast live events—not just film footage. Key engineers, including chairman of the system committee W. R. G. Baker (a GE employee), did not think their companies could "afford to complicate the black and white system by introducing on top of that, color." They worried color would be too expensive and would disrupt the introduction of black-and-white television. Had CBS done a better job gaining allies from among the large manufacturers before it presented its system to the committee, it might have had a better chance of authorization before the war. During 1941, only Zenith and Stromberg-Carlson expressed an interest in working with CBS in developing color television. FCC chairman Fly's decision to champion color was probably the main reason Baker instructed the panels to give it serious consideration. In comments later deleted from the official minutes of one of the first meetings of the system committee, Baker argued that he would oppose consideration of CBS color "if it came to the National Television System Committee."[4]

Although the commission seemed to support the recommendation of the system committee that it should give CBS a six-month period to conduct field tests, it never formally established any plans to evaluate CBS color after the completion of tests. With the entry of the United States into World War II, the FCC became preoccupied with defense-related activities. But the war also put on hold further expansion of monochrome broadcasting. CBS used this delay as an opportunity to improve its system and develop strategies for gaining support for color television after the war.[5]

Management changes at CBS during the war led to a clearly defined commitment at the highest levels to color television. Before the war, CBS executives had been divided over the importance of television to the business's future in broadcasting. Vice president Paul Kesten had been television's strongest supporter. He was instrumental in convincing the head of CBS, William Paley, to finance Peter Goldmark's efforts to develop an alternative television system the company could use to compete with RCA and its subsidiary NBC. During the war, Kesten's influence in the company was greatly strengthened. Paley spent most of the war years in London with the U.S. Psychological Warfare Unit and delegated many of his responsibilities to Kesten. Kesten made color television a top priority and authorized staffing changes benefiting Goldmark and his engineering team.[6]

Since Goldmark and members of his staff spent 1943 working for the government in war-related research, Kesten's efforts to promote color television initially focused on cultivating the support of manufacturers and other organizations. As seen in chapter 3, during the war, key government planners—including FCC chairman Fly—expressed an interest in a higher-quality television system that would take advantage of wide channels available in the newly opened UHF spectrum. Kesten saw this as an opportunity to promote his company's interest in a television system different from the one recommended by the system committee; he instructed his staff to plan for a high-definition color system using 16 MHz channels. Whereas the 6 MHz color system could produce 325 lines per frame, this wide-band system would yield at least 525 lines.[7]

CBS did manage to convince some of the large manufacturers to take an interest in developing a new television system, but the company realized it would also need to gain public support in order to convince the FCC to authorize color standards. The Radio Technical Planning Board was established in 1943 to advise the commission about postwar standards, but CBS had good reason to fear that RCA, the major company in television broadcasting and manufacturing, would dominate the proceedings. Beginning in April 1944, in an effort to cultivate public support, CBS kicked off a "quality television" campaign, which tried to portray RCA and the other "video now" advocates as enemies of technological innovation and progress. Kesten wanted the public to back his proposal to have the FCC and the industry use the war years to prepare for a superior alternative to the standards developed by the system committee.[8]

CBS's commitment to technological development as a business strategy was evident in the amount of support given Goldmark and his engineering department. After returning from government service, Goldmark began work on the wide-band system already planned by Kesten. By the end of the war, he was well under way toward completing a prototype of the new system. He completed the prototype in September 1945 and gave a closed-circuit demonstration to CBS executives. The company conducted the first UHF transmission of the wide-band, high-definition color system on October 19. Like the prewar system, CBS's new scheme also used a color disc and the field-sequence technique, but with the wider channels it could transmit 1,025 lines per image instead of 325. Throughout the remainder of 1945 and into summer 1946, the company conducted hundreds of demonstrations to various groups outside of CBS, including the press and members of Congress as well as the public. CBS executive Frank Stanton used data-quantifying consumer response to the broadcasts to support the company's claims that the public desired color. Stanton played an increasingly important role in promoting CBS's color venture after executive vice president Kesten retired in August 1946.[9]

The FCC declined CBS's November 1945 petition for authorization of its color process and the lifting of experimental restrictions on the use of the UHF spectrum, the commissioners again requiring CBS to first develop the capacity for transmitting live broadcasts. In September 1946, after perfecting a new wide-band color camera and developing the capacity for live pickup, CBS renewed its petition to the FCC. The commission was now ready to evaluate CBS color and scheduled hearings to begin in December.[10]

The hiatus provided by the requirement that CBS had to develop live pickup before the commission would consider its wide-band system gave RCA a chance to work on its own color design. Company executives were still committed to delivering monochrome television to consumers before introducing color, but they wanted to be prepared for any contingency, including the possibility that the FCC might approve CBS's field-sequence color system. In the summer of 1945, RCA announced that one of its employees, G. L. Beers, had patented an alternative field-sequence system. In case the FCC ruled in favor of CBS, RCA hoped it would be in a position to avoid becoming a licensee of its arch rival. The company also worked on two other systems based on simultaneous scanning of the image in the three colors and presented the different designs to the press late in 1945. Executives decided to concentrate the company's resources on wide-band simultaneous scanning, a system RCA had first presented to the FCC in 1940 but that was not considered at the time because the commission was interested only in narrow-band (6 MHz) proposals.[11]

RCA's simultaneous-scanning system used three electron guns to bombard the picture tube with a green, blue, and red image. Engineers employed a complex electronic and optical registration system to focus the three-color fields over each other at the same time to form a single picture. Since the system did not use a color disc, RCA portrayed it as all-electronic. The company played up the "mechanical" nature of CBS's system and implied it was less advanced because it used obsolete technology; this was somewhat misleading, however, because engineers expected eventually to discard the color disc once they developed a tricolor picture tube. The field-sequence system was thus not necessarily inherently mechanical. RCA's simultaneous-scanning system transmitted 525 lines per image at a rate of 30 frames (or 60 interlaced fields) per second. CBS's system had better resolution, but the lower frame rate meant it would more likely produce a picture that seemed to flicker. Like CBS's invention, RCA's was a wide-band system. It transmitted the three different color images in adjacent transmissions within a 13.5 MHz channel and recombined them at the receiver using the three electron guns. Because the green signal had the same standards as the existing monochrome system, broadcasters could use it to transmit, in black-and-white, the color broadcasts to monochrome receivers equipped with a UHF converter. RCA thought the "compatibility" of its color system was a major advantage.[12]

After the preliminary demonstration in 1945, RCA engineers returned to the laboratory and spent the next year working to improve the simultaneous-scanning system. When the FCC announced in October

1946 that it would consider CBS's petition for approval of field-sequence color, RCA asked if it, too, could present its simultaneous-scanning system. The company did not expect the FCC to make a decision about approving standards covering simultaneous color, but it hoped to convince the commissioners that they needed to consider alternatives to CBS's system before they made a decision that might freeze technical progress. RCA executives were especially interested in delaying a decision about color; the longer it took the FCC to evaluate CBS's petition, the more time monochrome television would have to become entrenched. They also wanted to convince the commission to adopt a system compatible with existing monochrome receivers. According to Charles Jolliffe, executive vice president in charge of RCA Laboratories, the simultaneous-scanning system was superior because it could be "introduced without penalty to the existing service and without jeopardy to the investment of the public and broadcasters in black-and-white television."[13]

The FCC Evaluates CBS Wide-Band Color, 1946–1947

The communications commission conducted hearings evaluating CBS's request for authorization of its color system from December 1946 through February 1947. When CBS executives petitioned the FCC in September, they had hoped a decision could be made by the end of the year. But the added work involved in evaluating RCA's system as well as another exhibit submitted by DuMont helped delay the final ruling until March 1947.

After the commissioners toured the laboratories of the different companies, they arranged for the systems to be displayed together for the first time during the hearings at the end of January. In preparation for the comparative demonstrations, CBS made some important improvements to its system. The brightness or illumination of the color picture tended to be reduced when the system transmitted the image through the color filters. In order to minimize this potential problem, CBS engineers increased the scanning rate from 120 to 144 color fields per second. CBS's system gave the best performance at the hearings. DuMont engineers did not have a complete system to present to the commission—only a three-gun picture tube that they could use with color television. RCA's simultaneous-scanning system was thus the only serious alternative to CBS's field-sequence color. But RCA engineers admitted they still had to overcome some important problems, including difficulties with image registration. E. W. Engstrom emphasized repeatedly that the company was presenting "a laboratory demonstration and not a finished show." Unlike CBS, RCA also had not yet developed the capacity for live pickup. RCA was mainly interested in demonstrating to the commission that CBS's system was not necessarily on the cutting edge of technological development. They used the color demonstrations to try to convince the commission that simultaneous color was potentially superior and to create doubt by pointing out faults in CBS's presentation. Despite views to the contrary by other observers, RCA engineers claimed flicker and inadequate brightness were major problems in CBS's demonstration. They also complained that the picture color would "breakup" and smear when an observer moved his or her head back and forth quickly. RCA officials argued that the registration problem in their system could easily be solved; by contrast, they believed, CBS's problems were virtually insurmountable.[14]

In its March decision, the FCC denied CBS's color-television petition. The commissioners were impressed by the doubts raised by RCA and some of its allies, which included the dozens of companies that held licenses with RCA to manufacture video components. An internal commission document concluded that "it appears to be the consensus of the industry, both manufacturers and broadcasters, that considerably more developmental work is required before standards are set." The two trade associations representing the television broadcasters and manufacturers testified against CBS. Only Zenith and Cowles Broadcasting supported Columbia's petition. The engineering department of the Radio Manufacturers Association not only argued that CBS's system needed more development and testing but claimed that RCA's simultaneous-scanning design was "superior" because it was free from flicker, color fringing, and color breakup. The association's engineers also stressed the importance of choosing a system compatible with black-and-white television. "With the simultaneous system the black and white service may continue to grow," according to the industry engineers representing the manufacturers association: "When the color service is introduced, these two services can develop side by side without obsolescence." They believed the sequential system, by contrast, would "result in maximum obsolescence." Although CBS seemed to have a head start over RCA in developing and perfecting its system, the engineering department of the manufacturers association estimated that, for each of the two systems to reach the level of monochrome television in 1946, the difference in time needed was "not of a significant magnitude"—about four years for field-sequence and five years for simultaneous-scanning.[15]

The official report argued that the FCC was not convinced by the "evidence" that CBS's color system represented "the optimum performance which may be expected of a color-television system in the forseeable future." The report contended "that there may be a number of other systems of transmitting color which offer the possibility of cheaper receivers and narrower band widths that have not yet been fully explored." The commission was particularly critical of the mechanical nature of CBS's system. It did not believe "a mechanical filter at the receiver would be accepted by the public." Staff members were concerned that "there is no assurance that the mechanical filter can be replaced by an electronic one." The commissioners were not prepared to tolerate much uncertainty when evaluating CBS color. They demanded a high level of field testing before granting commercial operations. The staff emphasized the importance of providing "a healthy safety factor so that plenty of latitude will be available for non-optimum viewing conditions as well as future non-predictable exigencies in the art." As noted in chapter 5, commissioners were partly concerned about a lack of understanding of the propagation properties of the UHF spectrum, but they also did not think CBS had conducted adequate tests of color receivers and transmitters. They were concerned that the public should not be treated as a "'guinea pig' for color television."[16]

The commission's low tolerance for technical uncertainty partly reflected the critical view of the role of technical evaluation in policy making in the postwar period (see chapter 5), but it is also important to acknowledge that the FCC tended to emphasize this position most strongly when it did not conflict with the wishes of the dominant interests in the industry.

The commissioners may have been genuinely convinced by the doubts raised by opponents of CBS color, but many of the commissioners who were appointed after Fly resigned in 1944—as well as their support staff—generally were not inclined to oppose the dominant broadcasters and manufacturers. An internal document connected with the March 1947 report argued that "standardization for an industry as wide spread as television requires the whole hearted support of the industry involved." CBS was not only virtually alone in pushing field-sequence color, but unlike RCA it did not have manufacturing capabilities to support its broadcasting operations. One concerned staff member pointed out that the commission was "considering a standardization proposal from a party who will only be secondarily interested in reducing the system to practicality." The risk to the continued growth of monochrome television did not seem to justify a positive decision about CBS color based on uncertain knowledge. A handwritten note by a commission staff member contended that "an acceptable black and white system is in operation today. The proposed system would provide no better definition, no better coverage, no better programs. It would provide color. . . . The risk of overturning the whole television cart is lurking in the shadows." The FCC's apparent preoccupation with supporting the status quo led CBS to accuse the chairman, Charles Denny, with improper conduct when—six months after the color decision—he resigned to accept a high-level position at NBC. But even if Denny had been guilty of giving RCA preferential treatment in exchange for employment, there are no records to indicate that anyone else on the commission, including members of the engineering staff, opposed the decision against CBS. It would break the bounds of credulity to suggest that all the commissioners and staff could have been guilty of gross improprieties.[17]

Like other decisions involving technical issues, the FCC ideally wanted its rulings on CBS color to be based on "sound scientific" evaluation. However, as we saw in previous chapters, the decision-making process often involved a complex resolution of tensions among different kinds of issues.[18] With color television, we see new complexities involving contested meanings of scientific evaluation. On the one hand, policy makers wanted technical experts to evaluate the quality of CBS color; on the other hand, they also contended that scientific evaluation should involve extensive field testing under home conditions using nonexpert observers.

While the Radio Technical Planning Board and the Radio Manufacturers Association mainly used radio and television engineers employed by manufacturers and broadcasters to evaluate the different color systems, the FCC also asked at least one academic physiologist, or vision expert, Selig Hecht, a professor of biophysics and head of the biophysics laboratory at Columbia University, to attend the demonstrations and testify at the official hearings. Although the commissioners never stated this explicitly, they probably assumed that since he was an independent expert, his testimony would be more trustworthy and credible than that of the corporate engineers. Hecht's testimony covered over one hundred pages of the official *Proceedings*. During the war, he had been a member of a number of special committees established to assist the U.S. Army and Navy with technical problems involving "the physiology, chemistry and physics of vision." He also had helped evaluate standards for black-and-white television as a member of the NTSC. Hecht generally disagreed with RCA's criticisms of CBS color. "To judge as an experienced visual observer," he testified, "I should say that CBS has produced an acceptable color television picture adequate in brightness, color, resolution, contrast and freedom from intrusive flicker."[19]

But Hecht's testimony also underscored the complexities involved in scientifically evaluating color television. Hecht pointed out that because a number of "conflicting factors" were involved in producing an acceptable television picture, "a compromise" had to be made "so that the broadcasting of television can proceed on an acceptable basis." For example, flicker, brightness, frame rate, and channel width were closely related. Given the same frame rate, flicker becomes more likely as the brightness is increased. If experimenters increase the frame rate, they can raise the brightness level to a higher value before flicker occurs. But if they increase the frame rate, they then have to expand the channel width; this, in turn, limits the number of stations, nationwide, that planners can authorize. A final decision about what constitutes adequate brightness or acceptable flicker thus depends on an accommodation among these different factors. "This business of saying what will flicker and what will not flicker," according to Hecht, "is not so easily stated because a good many facts enter into the situation."[20]

Hecht also pointed out that subjective or psychological qualities of individual observers would help determine how they evaluated such issues as flicker and brightness. He testified that in his own observations of CBS color, he did not notice flicker until he "started to look for it with great care, putting myself in such a position, fixing my eyes very carefully so that my head would not move and therefore get the flicker best." He also admitted that different observers would evaluate the same television picture differently. For example, based on a study he had conducted with monochrome television for the Public Health Service, he estimated that 2 percent of all observers would see flicker even at one-fifth the brightness level at which the average person would observe it.[21]

A recognition that planners would need to base television standards on extensive testing using many different kinds of observers led to demands that large groups of people should be studied. This supported RCA's argument that CBS should conduct more field testing of its system under household conditions. According to an RCA engineer, since experts admitted that in evaluating color-television standards they were "dealing with subjective matters" and "cannot make objective measurements accurately," they should be prudent and cautious and take into account "persons whose acuity or whose conditions of observation differ from the average." CBS had not conducted true "scientific tests," according to RCA and other critics, because they had not randomly chosen a large number of untrained observers to test their system; they relied too heavily on trained experts and CBS employees.[22]

The FCC supported this view that it should not consider "the establishment of standards until color is out of the laboratory and has been in the hands of practical non-engineering people for a period of months." "The brightness and flicker problem," according to one handwritten memorandum, "appears to be too subjective for engineering analysis." Some of Hecht's testimony actually favored the position that trained experts were out of touch with the practical realities of home viewing. Hecht admitted that he did not own a television set and did not regularly view monochrome television broadcasts.[23]

Unlike when it was under Fly's leadership, Denny's commission was not especially interested in exploring the relationship between the technical advice it was receiving from industry engineers and the economic interests of the engineers' employers. For example, the commission did not seem to be very concerned about how inconsistencies in engineers' testimony might be related to changing corporate strategies. Craven, who had served on the commission when it authorized standards for monochrome television, argued in December 1946 that the same companies that had "begged the commission to set standards for black and white television in the lower bands and blamed the commission bitterly for delaying the development of television" are now "pleaders for delay when a competitive system is just as ready as low-band monochrome television was before the war." The commission did not publicly explore the underlying economic or business motivations that might be guiding engineering testimony. Craven's own testimony might also have inspired commissioners to deal more explicitly with this problem of advocacy versus objectivity. Before the war, Craven had at first criticized Fly's efforts to prevent RCA from freezing monochrome standards and from gaining a monopoly over television based on control of key patents. He was not strongly in favor of the FCC basing decisions on patent policy and the potential for monopolistic domination. But once he became involved with Cowles Broadcasting, which was supporting CBS color as a way to compete with RCA and the major broadcasters and manufacturers, Craven contended that the commission should authorize field sequential because it would favor the "smaller pioneers" and not lead to monopoly control.[24]

Although in the official reports the commission did not explicitly deal with issues related to the tension between advocacy and objectivity, there is some archival evidence that staff members were looking for ways to explore this problem. One memorandum analyzed the different "motives of CBS and other principal parties for presenting testimony." The discussion focused specifically on how the positions of the different companies with respect to color-television standardization might be influenced by the patents they held. The staff member who wrote the memorandum informed the commission of the "thirteen patents which relate to color television" held by Columbia, including "one of the patents . . . fairly basic to the sequential system."[25]

According to the report, the company would clearly benefit financially from authorization of its system since other companies would have to purchase licenses to manufacture equipment based on the system. The staff member also stressed that RCA had a strong economic motivation to promote the growth of VHF monochrome television: "Under the transmission standards adopted by the FCC" for black-and-white television, "RCA has the field very well covered by its patents." RCA authorized more than seventy-five licenses to other companies to manufacture and sell television receivers. Although RCA was anxious to "move forward and recoup some of its great investment in black and white television," according to the memorandum, it was not "ready to go forward with color television from a patent viewpoint." If the FCC authorized CBS's color system, other major manufacturers—including DuMont, Philco, Farnsworth, and GE—would "be required to also take a license under the CBS patents," in addition to paying royalties to RCA. The staff member pointed out this would substantially "cut down" their profits.[26] Significantly, the report expressed doubts about the recommendations of the Radio Manufacturers Association: "It is difficult to know just what weight can be given to its recommendations for the reason that there is always a possibility this organization is more or less controlled by a few of [the] larger radio manufacturers such as RCA, GE and [a] few other companies who are RCA patent licensees." Although the staff member wrote this report to guide commission decision making, there is no evidence that commissioners took into account the doubts and critical questions revealed in the memorandum.[27]

Another staff member elaborated not only on the proper role of the FCC in directing color-television development, but also on the proper role of technical experts in advising the commission. This proposal argued that the commission "must take the reins and actively direct the experimentation and research," partly by organizing a new engineering advisory group "to be known as the Color Control Board." According to the proposed plans—which sounded very similar to Fly's earlier efforts—the new organization would "supercede" the Radio Manufacturers Association and the Radio Technical Planning Board, because "Columbia [CBS] and others" have "lost confidence" in these two groups and "would not cooperate." Significantly, the staff member believed "more emphasis must be placed on selecting impartial scientists that are guided only by engineering considerations." The proposed "Color Control Board" would be "actively dominated" by representatives of the FCC: "We will select the other members and balance the weight of different schools of thought. . . . More variety must be secured on the committees, small experimenters and undominated licensees must be appointed."[28] The proposal was critical of the engineers who claimed they were putting on different "hats" when testifying before the commission as both company representatives and members of engineering advisory committees. The proposal did not identify where to place the boundary between "engineering considerations" and other factors, but the staffer did emphasize the importance of establishing proper institutional arrangements to help solve some of the dilemmas related to the tension between advocacy and objectivity.

The commission never formally followed through with this or other proposals to plan for authorization of color standards, partly because commissioners were more interested in supporting the status quo but also partly because CBS discontinued further development of its system. The March 1947 decision shocked CBS executives; they had expected the FCC to rule in their favor. Discouraged and frustrated, CBS officials decided in April to halt research, although the commission had not ruled that CBS's system was substandard. For the most part, the commissioners did not give their own evaluations of the demonstrations; they mainly responded to the doubts raised by critics. But CBS still had the impression that the FCC was ready to side with RCA and its supporters.[29]

CBS decided to discontinue color development before they lost out completely in the race for monochrome television. The company had made a decision in 1945 not to apply for VHF stations. By 1947, the other networks had taken many of the VHF channels available in the major markets. CBS had to scramble to build a network of black-and-white stations. Network affiliates were strongly pressuring the company to devote all its resources toward monochrome broadcasting. CBS would have found it difficult to continue color development since its allies among the manufacturers—Zenith, Remington Rand, and Bendix—decided not to support production of color components. Cowles Broadcasting also opted to devote its resources exclusively to monochrome.[30]

Instead of working on color television, Goldmark's research team pursued a broad range of alternative projects during 1947 and much of 1948. Company executives supported Goldmark during this period mainly because of his work alongside engineers from the Columbia Record Company in the development of the long-playing (33 1/3 rpm) record. Goldmark's success with this project helped convince CBS executives to allow him to resume work on color television. They refused to provide further financial assistance, but they did allow Goldmark to work on color television with the assistance of outside sponsors. As early as the 1930s, medical schools and equipment manufacturers had expressed an interest in using a color system to televise surgical operations. Color seemed especially necessary for representing, in great detail, all aspects of human organs and anatomy. At least one medical equipment firm had observed CBS's color demonstrations during 1946 with the idea of producing a system to sell to medical schools. Goldmark was able to tap into this interest and, in the summer of 1948, convince the same company (Smith, Kline, and French) to sponsor further research.[31]

A New Commission Champions Color Television, 1949

The "television freeze" went into effect soon after Goldmark resumed color research at CBS. The commission wanted to use the freeze to reexamine all issues related to the establishment of a nationwide television system, including color television and UHF as well as the assignment of VHF stations (see chapter 5). The FCC felt pressured to deal with the color issue first, mainly because of intense lobbying by Senator Edwin Johnson, the Democratic chairman of the Interstate and Foreign Commerce Committee. Johnson had become interested in issues connected to the freeze partly in response to the complaints of Eugene McDonald, the head of Zenith, who claimed that RCA and other manufacturers were not giving the public sufficient notice about possible obsolescence of VHF receivers. Johnson was concerned that the "commission policies in television had been such as to foreclose, or at least slow up, commercial development" of color television. He wanted to make sure the FCC did not ignore opportunities to prevent "monopoly control . . . in the manufacture of the equipments used for transmission and reception of television." Both RCA and the FCC, according to the senator, needed to answer to the charge of "RCA's alleged undue influence in commission's policies." In criticizing RCA, he expressed skepticism about the reliability of the testimony of one of the company's foremost engineers, Charles Jolliffe. "Jolliffe thinks that his employer is the greatest and finest in the world," Johnson mocked. He has "that tight engineering mind which sees things only one way—his way; whoever differs from him is incorrect. . . . I have been in public life too long to take too much at face value." Johnson specifically criticized Jolliffe for acting as if his engineering expertise made him an expert in "the field of analyzing regulatory practices."[32]

Johnson expressed his concerns to the commission in a letter sent in February 1949 and in a speech in the Senate two months later. Although theoretically government officials had established the FCC to be independent of the major branches of government, in reality Congress can pressure the commission if it strongly disapproves of its activities. Congress controls the funding and can refuse to approve the president's appointments. Since Fly had been forced to resign because he lost favor with Congress, the commission not surprisingly responded promptly to the senator's attack and called for hearings on color television to begin in the fall. Companies were expected to present proposals for their own color systems at these hearings. The commission assumed that it could deal with the color issue reasonably quickly, allowing more time to solve the other problems that had led to the freeze. As we will see, the pressure exerted by Congress helped push the commission to overcome its traditional tendency to avoid making a decision that might go against the dominant factions in the broadcasting industry.[33]

The FCC's response to Johnson's letter of February 1949 pointed to a new willingness to consider the economic structure of the industry in deciding about standards in order to prevent companies from gaining a monopoly based on patents. The legal justification for these actions had not been entirely clear to earlier commissions. A strict interpretation of the monopoly provision of the 1934 Communications Act seemed to authorize the FCC to act against broadcasters only when they had previously been convicted of violating the antitrust laws. Chairman Wayne Coy argued that a recent Supreme Court ruling legitimated a broader interpretation, specifically authorizing the commission to act against licensees based on present violations. Coy acknowledged that the Communications Act did not authorize the FCC to take action against manufacturers directly, but he believed the court ruling also clearly gave the FCC legal authority to act against manufacturers who were also licensees of stations and who used their patents to acquire monopoly control. According to this view, the FCC could deny licenses to stations controlled by manufacturers who exercised "monopolistic patent control . . . or any activities which constitute restraint of trade or unfair competition within the meaning of the Sherman or Clayton Act." Coy assured Senator Johnson that the FCC was already in the process of "conducting a study of the patent situation in the radio field and also the practice of companies in buying patents which they do not themselves own for licensing to others in order to determine whether such practices are inconsistent with the Sherman Act."[34]

Although Senator Johnson played a crucial role in the FCC's decision to conduct a thorough, wide-ranging, and critical evaluation of color television, the influence of new commissioners, including Coy and especially Robert F. Jones, also needs to be acknowledged. Only two of the commissioners who had participated in the 1947 hearings were involved in the 1949–50 deliberations. Partly because they did not have a personal stake in the earlier decisions, the new commissioners were more open to reexamining postwar decisions made by previous commissioners. Before he was appointed to the commission, chairman Coy had actively supported the efforts of the FM broadcasters trade association and had worked with Armstrong and the other FM enthusiasts who believed RCA and the dominant AM radio interests were seeking to stifle the new technology. He was accordingly more likely to be sympathetic to CBS's efforts to promote its color system against RCA resistance. Jones was the strongest promoter of color on the commission and supported Senator Johnson's attempt to link the issue to the traditional worry about possible monopolistic dominance by RCA of radio and television broadcasting and manufacturing.

A number of the other new commissioners had also served as staff members on earlier commissions. George Sterling, for example, had previously been the FCC's chief engineer. In some cases, familiarity with the earlier decisions—and their consequences—probably provided an incentive for a reexamination of all the issues connected with the "freeze." Some of the views of the staff members involved in the 1947 decision had apparently been disregarded by the commissioners in 1947 (see the above discussion of the memoranda written by support staff who recommended, among other things, that the FCC should take an active role in planning the development of color).

In preparation for the color hearings scheduled to begin in September 1949, Goldmark, CBS's chief engineer, perfected a revised version of his earlier system. The new color system had a number of improvements, including, most importantly, the ability to use channels with a bandwidth of 6 MHz. It produced 405 lines per frame and operated at 144 interlaced color fields per second. Goldmark enhanced the resolution or picture detail partly through the use of a new technique known as "crispening," which enhanced the transition between the light and dark edges of objects. Columbia demonstrated this 6 MHz system to commissioners and staff during October and November 1948 and to executives from his sponsor, Smith, Kline, and French, during the next few months. In June 1949, Goldmark used his color system to televise surgical operations at the University of Pennsylvania Medical School; he also gave a demonstration at a meeting of the American Medical Association in Atlantic City, New Jersey. When, in 1947, the FCC had refused to approve CBS's system, it had emphasized that the company had not done enough field testing. Columbia thus sought to complete as much field testing of its improved system as possible before the new hearings in 1949. Experimental broadcasts from WCBS in New York began in July of that year, and in September from WOIC in Washington, D.C. Westinghouse and Remington Rand agreed to manufacture equipment for CBS, and this included receivers that the company distributed to as many different kinds of viewers as possible.[35]

Despite RCA's assurances during the 1947 hearings that its simultaneous-scanning system would soon be perfected, difficulties with image registration and other problems forced the company to prepare a new system to present to the FCC for approval. For the commission's hearings in the fall of 1949, officials rushed to develop a new apparatus using a method called "dot sequential." Although it still needed further testing and refinement, RCA's new system was the only major rival to the system CBS was presenting. A small company called Color Television, Inc., also presented a "line sequential" system at the hearings. The commission took the company's proposal seriously, but at no time did the system present pictures that could rival those of CBS.

Instead of alternating color fields, RCA's dot-sequential system alternated color on picture elements; Color Television Inc.'s line-sequential system alternated color on different lines. Since RCA's new system, unlike CBS's, operated on 525 lines per frame and 30 frames per second, it was essentially compatible with the established monochrome standard. However, it was also more complex. Each picture line consisted of color dots, in the sequence red, blue, green. The first field scanned the dots in the "odd" lines in a sequence beginning with the odd-numbered dots in the first line and the even-numbered dots in the third line. The second field scanned the "even" lines, following a sequence beginning with the even-numbered dots in the second line and the odd-numbered dots in the fourth line. The third field scanned the remaining sequence of alternating dots in the odd lines and the fourth field scanned the previously unscanned dots in the even lines.[36]

The camera in the RCA system used three camera tubes each set up to receive one of the primary colors through the use of color-selective mirrors. Instead of using three cameras to scan in all three colors simultaneously, the system used the cameras in a sequential manner, one following the other down each of the lines. Together, the scanning information from the three guns, when combined, would not fit into the 6 MHz channels. To solve this problem, RCA introduced an approach called "mixed highs," which took advantage of the observation that the human eye does not distinguish fine detail in color. Coarse detail for each color was transmitted in the lower half of the 4 MHz available for the video channel; the system mixed the fine detail and then transmitted the information in the upper-frequency band as tones of gray. To save more bandwidth, the apparatus passed the three-color signals through special sampling circuits to filter out the low frequencies. Equipment combined the remaining high frequency components as mixed highs. Broadcasters could then transmit all the information to color receivers equipped with mirrors and color tubes. The system had to carefully synchronize the images from the color tubes to assure quality pictures. When it functioned correctly the dot-sequential system produced higher-quality color pictures than the simultaneous system, but, like the simultaneous system, RCA's new system was also very sensitive to registration problems.

The FCC's hearings on color that began in September 1949 ended in May 1950. The commission took a three-month break in the middle of this period to give companies an opportunity to conduct field tests and improve performance. Using live broadcasts from its station in New York City, CBS completed tests with receivers located in at least twenty-five homes throughout the city. After further tests in other cities—St. Louis, Washington, D.C., and Baltimore—the company believed it had satisfied the FCC's 1947 requirements for field testing. All three companies demonstrated their systems to the commission during both phases of the hearings. The commission held comparative demonstrations of CBS's and RCA's systems in both November and February. Color Television, Inc., was not ready to present its system until the second comparative demonstration in February.[37]

Technical Experts and Policy Making on Standards

The main objective of the FCC at the 1949–50 color hearings was to decide which of the three systems was superior and deserved authorization as the national standard. As in the earlier hearings, the commission tried to use engineering advisory groups to provide expert objective evaluation. They wanted to use organizations that "will furnish the commission . . . with objective testimony which will look down upon the private considerations or selfish considerations of the several manufacturers who might have a profit interest at stake." The Joint Technical Advisory Committee was the main group that professed to fit this definition. As we saw in chapter 5, the committee ignored its members' corporate affiliations and claimed to be interested only in "the public interest," which it equated with providing "sound technical determinations." But the assistance of the joint committee as well as the Radio Manufacturers Association again points out the complexities involved in "scientifically" or "objectively" analyzing different systems. Key commissioners, especially Robert Jones, questioned the value of this expert advice by stressing how the experts did not adequately deal with these complexities.[38]

Jones expressed his criticisms of the joint committee when he questioned its representative, Donald Fink, who was invited to testify as one of the first witnesses at the color hearings. Jones thought the committee overidealized the problem of choosing a color system by analyzing systems that were theoretically possible, instead of evaluating actual demonstrations of functioning apparatus. The committee investigated eight possible systems (using both 6 and 12 MHz channels), including three kinds of "simultaneous" systems and five types of "sequential." Jones accused the engineers of adding "several systems, merely papers systems, in order to build up straw men to knock down existing systems." He thought the engineers on the joint committee were too closely allied with the manufacturers association and he did not think it was right that broadcast engineers heavily involved in one industry (black-and-white television) should be responsible for evaluating a new, competing industry—color. According to Commissioner Jones, it did "not seem fair in the principles and concepts of American business or American justice that a competitive system cannot come forward until it has passed the judgment of those who would compete with that new industry." The executive officers of the manufacturers association had to approve the membership of the joint committee; Jones believed that these members mainly represented RCA and the other companies that had an economic interest in monochrome broadcasting.[39]

The joint committee evaluated the theoretical performance of the different systems based on such characteristics as resolution, color fidelity, and flicker. These evaluations were placed in the class A category—the category for data considered to be the most credible (see chapter 5). However, Fink admitted that these evaluations were purely qualitative. Had the commission asked for quantitative comparisons, he testified, the committee would have had to place most of the information in the class C category—that for the least-reliable data. Jones and other commissioners questioned the practical value of the evaluations of the joint committee, not only because they were purely theoretical and did not take into account the economic costs of different systems (just the "technical" characteristics), but also because, as Fink admitted, the committee did not quantitatively study actual, functioning, systems "on an apparatus standpoint" since the members "didn't have the time for the further work." Potentially more disturbing to Jones and other commissioners was Fink's confession that he had a "prejudice against moving mechanical parts of high speed in a receiver." Since the system that CBS was demonstrating still relied on a mechanical disk, or drum, to impart color to pictures, Fink's admission seemed to indicate that he had a preconceived bias against the CBS system.[40]

When questioned as a private citizen, rather than as a representative of the joint committee, Fink agreed with Jones's contention that the FCC should base its color decision on extensive scientific field testing of the different systems. Fink advocated using the kind of "controlled tests" of viewer responses to different color receivers that the medical profession used when evaluating drugs. "This must be a controlled experiment," he said, "not one in which the market place and all the things that go with the market place, determine what are the technical facts"; experts were needed to establish the testing structure objectively and then evaluate the subjective reactions of different consumers.[41]

The commissioners had briefly analyzed the problem of how to evaluate proper field testing during the 1947 hearings. In order to analyze the issue more thoroughly during the new hearings, they asked an FCC staff member who was an expert on scientific sampling to respond to questioning as a witness. H. H. Goldin's official title was acting chief of the Bureau of Accounting's division for economics and statistics. The commission was particularly interested in asking Goldin about the possibility of using a "scientific audience survey" as a definitive guide for judging which system was superior. This approach seemed ideal for finessing the tension between two important demands on decision making emphasized in this book. It would satisfy both the democratic demand for policy based on nonexpert involvement and the perceived need to rationalize considerations based on rigorous technique.[42]

But Goldin emphasized the difficulties involved in trying to set up a "scientific ideal." He did not think the FCC had enough time or sufficient resources to complete such a survey. It would take at least one year, he argued, to gather a proper data set. The staff would first have to take a limited region and conduct pretesting in order to clarify the questions that it would need to answer. After conducting trial surveys, they would have to meet with other experts and decide if further trials were needed to remove statistical biases before completing the final study. Even if there was time to complete all these steps, Goldin warned, the testers probably would not be able to translate public reactions into useful engineering standards. He thought it would be very difficult to ask average viewers "technical questions on engineering aspects" such as resolution and brightness, when consumers did not know what those terms meant in any precise sense.[43]

Goldin also criticized other surveys conducted by the companies introducing color, including an extensive study conducted by CBS that tallied the responses of more than twelve thousand individuals who watched CBS color presentations and were then asked about such issues as brightness and flicker. He contended that the CBS survey—despite its large data set—did not attempt to put together a "scientific sample" and did not try to remove all biases. It did not, for example, explore what would happen if people watched color more than once. Would they "have the same response the tenth time as they did the first time?" he asked. "There are certain rules which exist in this," according to Goldin, "certain basic knowledge, and on the basis of that we can take these rules and test against a particular result which is shown." Goldin further pointed out in response to questioning that the limited surveys done in preparation for the monochrome decision in 1941, which mostly involved counting letters and postcards from viewers, had no validity "from a technical point of view." Underscoring the difficulties involved in conducting a scientific survey of consumer reaction to color, Goldin went so far as to indicate that it might be nearly impossible; he did not think "any survey could be made which would stand up scientifically in terms of some of the issues involved here."[44]

When Commissioner Jones asked how the commission should evaluate and use the different public surveys that seemed not to have scientific validity, Goldin responded that the commissioners would have to use their own judgment in determining the worth of the surveys as well as that of expert advice. He claimed that personal experience with the different color receivers, which "would give them some information which they might not get from a technical source," would be especially important. Goldin argued that the commissioners "should have the opportunity to have the sets in their homes and to play with them for some time, and they will have the advantage of working the dials themselves and they will have the advantage of how does their wife react to it, and how do their children react to it."[45]

Jones, in choosing among the different sources of information that might be used to help formulate policy—the choices included engineers, consumers, and the commissioners, with their families, themselves—favored placing the most emphasis on consumer response. He thought the engineers were too closely "connected with the industry" and were more likely to "have one reason for bias or another." This idea that industry engineers could not be trusted was also expressed by a representative of Arco Electronics, a company that made radio kits for hobbyists. The company wanted the commission to "rely for guidance on the judgment and experience of 250,000 or more, experimenters, who are neither naive laymen, nor biased and hypersensitive engineers." The company representative believed that since FCC and industry engineers admitted that "all systems involve technical shortcomings which result in unacceptable products," further experimentation was necessary. If the commission would order broadcasters to provide color telecasts of the different systems, the company would supply "all essential parts or components . . . and instructions . . . illustrating in simple manner to the experimenter how to build a unit." To this Arco Electronics representative, the use of trial-and-error experimentation by hobbyists seemed ideal, not only because it would take advantage of "unbiased and impartial minds emanating from every section of our country" but also because it would "be in full conformity with our fundamental American concept of free enterprise." Although opponents could argue that this proposal was merely self-serving, it does illustrate the importance of a traditional populist theme—one that Commissioner Jones, too, was engaging.[46]

During the hearings, the commission questioned the major manufacturers about the availability and potential costs of equipment for the different systems. A number of the companies refused to cooperate. DuMont and Philco declined a request to discuss their manufacturing plans with Jones in preparation for the hearings. David Sarnoff of RCA alienated the commissioner by implying that Jones would not be able to understand the "complicated technical details" of his company's plans. Jones was particularly interested in finding out if the manufacturers would cooperate if the FCC decided to adopt the CBS color system or a flexible system that would accept both standards for CBS color and other color systems compatible with existing black-and-white receivers. The idea for a flexible system was similar to a proposal DuMont had made in 1940 in preparation for monochrome television, for standards that would accept anywhere from 400 to 800 lines and frame rates from 15 to 30. These multiple standards had not been adopted, but Jones reminded industry witnesses that the final standards for monochrome television had included multiple synchronization options, including standards originally proposed by Philco, Hazeltine, DuMont, and the manufacturers association. He thought this example provided an important precedent for developing a flexible, or at least composite, system in color.[47]

The industry response to the commission's inquiries about possibly supporting a decision adopting CBS color or flexible standards was mixed. The manufacturers association did not specifically agree that it would support CBS's system if it was adopted, but the group did concede that it had a legitimate duty to cooperate with the commission in helping to implement authorized standards. Most companies in the industry remained hostile to CBS's efforts to promote its color system; they also did not like the idea of having multiple standards that would include CBS's system. Philco argued during the hearings in April 1950 that business growth would be possible only with "a single set of specific and detailed standards." The company representative made a prediction that might have been a veiled threat when he contended that if the commission authorized multiple standards, "I suspect that what would happen is that the industry would get together tacitly or otherwise and arrive at a preferred system and go ahead and make it." Along similar lines, a representative of Color Television, Inc., predicted that flexible standards would not result in different systems fairly competing on a level playing field. So long as RCA maintained a dominant patent position in the radio field and a dominant role in manufacturing and broadcasting, he argued, "just so long will so-called competition in the market place under any adopted multiple standards which include the RCA system be a vain and an impossible mirage."[48]

We gain a deeper understanding of these 1949–50 color hearings by analyzing the debates in the context of the tension between technocratic and nontechnocratic views about the role of engineers and policy-making institutions in decision making on standards. The FCC commissioners believed their institution, rather than industry technical groups, should play the most important role in deciding color standards. Commissioner Jones made sure the commission questioned industry representatives closely about patent considerations and other "nontechnical" questions. For example, he wanted to know if RCA was illegitimately trying to use its patent licensing and sublicensing policies to gain industry dominance. CBS agreed that broad policy questions, which were the responsibility of the commission to solve, should be thoroughly explored and should not become "lost in the maze of conflicting technical data." The company wanted the commission "to shift the emphasis from the myriad of technical data—where it has been in the past—to the handful of broad policy questions and issues which define the problem of color television." This preoccupation with "technical details," according to a CBS representative, was one of the main reasons the public still did not have color. Technical details were important, but they needed to be placed in their proper relationship to broader issues. "If the basic policy issues are considered first—and are accurately defined," he believed, then "the various technical data may perhaps be viewed more readily in their proper relationship to the problem as a whole." Important policy questions listed by CBS included whether it was in the public interest to establish a color system promptly and what criteria should be used to evaluate the different systems.[49]

RCA engineers and other industry representatives rejected the views of CBS and the commission by expressing an extreme technocratic position. RCA engineer Elmer Engstrom believed that when the FCC authorized color standards, it should not take into account "irrelevant" considerations such as patent policy and questions about who controls particular technological developments. "As a person interested in the technical matters concerning the starting of a new service there is only one criterion that ought to be used if we are looking in terms of public service," Engstrom argued, "and that is that the performance be the best we now know how to make it, regardless of the building blocks that are used to provide the service." Since television standardization was primarily a technical problem, following its own internal logic of development, he believed that performance could be objectively analyzed by engineers or other technical experts. Engstrom thus closely identified the public interest with technical evaluation. The president of RCA, David Sarnoff, expressed a similar technocratic faith in industry engineers or "scientists" to solve any problems that might arise in developing color-television standards. He testified that "you have got to have faith in your scientists, that they are going to be able to cure these difficulties as you meet them."[50]

This technocratic faith in experts and engineering evaluation was especially important when industry engineers reestablished the National Television System Committee in January 1950 and tried to convince the commission to give it the same authority it had had when monochrome standards were developed. As in 1940, GE's Baker was again chosen to serve as president. He did not think the three different systems being considered by the commission represented the best combination of standards that were possible. He specifically testified in spring 1950 that if he had to select a system, he would not choose RCA's. Baker wanted the commission to submit to industry engineers serving on the system committee all the evidence it collected and let them make a final decision about the best set of standards. Like the first system committee, the commission would supervise this group, but he did not think the FCC—or for that matter other existing advisory groups, including the Joint Technical Advisory Committee—had adequate time, resources, or organization actually to conduct the necessary study.[51]

In justifying a preeminent role for the system committee, Baker argued that the work of the 1940 committee demonstrated that industry engineers could use reason and empirical evaluation to "come down to a unanimous opinion." Baker presented a decidedly idealistic view when he claimed that the members of the first system committee had not been under "pressure patent-wise or commercial-wise" or that the commission or any industry faction had used "force" in getting members to agree about standards. He believed that "if you get these engineers together as engineers . . . and put all the facts and figures out on a piece of paper out on the table away from the influence of any commercial considerations or urgency, beyond the point where you want to get the job done within a reasonable time, I have never seen the case where you could not get an answer." The engineers serving on the system committee would ignore commercial considerations, he thought, because they would not be evaluating the merits of systems presented by particular companies but would be trying to establish an ideal, composite system based on "a terrific mass of information."[52]

Partly because it believed Baker was presenting an idealized view of the system committee, the commission was skeptical of his testimony. The commissioners did not think the industry was truly committed to color television and they were not convinced that industry engineers who had developed particular color systems would be able to provide an "objective evaluation." The commission counsel asked: "Could you ever divorce Peter Goldmark from the field sequential system, divorce Dr. Engstrom from the RCA system?"[53] Baker also lost credibility with the commission when embarrassing inconsistencies were revealed during cross-examination. Although Baker claimed the system committee should not make decisions about such issues as whether a compatible system ought to be chosen, an FCC representative pointed out that one of the committee's panels had already made such a decision one month earlier in favor of compatibility. Baker argued that decisions of the individual panels did not necessarily mean very much, but this revelation seemed to indicate that individual members had their own agendas, which included their own views about the relationship between the commission and the system committee. Baker was further embarrassed when he acknowledged that key passages had been deleted from the minutes of the meetings of the first system committee. His tendency to make offhand, flippant remarks also tended to undermine his position with the commission. For example, in 1940 Baker told someone that "the main committee [of the system committee] is feeble-minded; recommend anything you want." Referring to this comment during the FCC's hearings in 1950, a CBS representative asked Baker sarcastically, "Do you think that a committee which is feeble-minded is better qualified to choose ultimate standards than the commission?"[54]

The commission's rough treatment of Baker and rejection of the system committee's offer for assistance helped further alienate industry representatives, which made it more difficult for the FCC to convince the industry to cooperate with its decisions. Jones was particularly harsh in his questioning of Baker. When Baker admitted that he was not prepared to respond to questions about the 1940 monochrome decision because he had been too tired to look over the material during the evening before he testified, Jones responded that "if you are not interested in helping the commission out in the public interest, I suppose we could all go to bed and go home and call the whole hearing off, but we want color—at least I want color."[55]

After the conclusion of the FCC's color hearings in May 1950, the commission spent the summer analyzing the evidence before making a final decision about standards. The record available for the commission to study also included the results of an important year-long inquiry initiated by Senator Johnson in May 1949. A five-member advisory group organized and chaired by the director of the Bureau of Standards, E. U. Condon, conducted the study. Johnson had been interested in obtaining an "independent appraisal of the present status of color television" for his Senate Committee on Interstate and Foreign Commerce. He wanted "sound, impartial, scientific advice" from a committee composed of "scientific persons of repute, none of whom are employed by or have any connection directly with any radio licensee or radio-equipment manufacturer." Because Johnson thought the FCC had delayed the introduction of color television "for reasons difficult for us to understand," he also thought the advisory group should be independent of the FCC. It seemed appropriate to turn to the Bureau of Standards for assistance since the bureau's engineers had been involved in radio research since the first efforts at regulation; they seemed independent of influence from either industry or the FCC.[56]

Like other individuals involved in broadcast policy whom we have analyzed, Johnson wanted a rigorous, scientific study that would circumvent conflict and provide closure by rationalizing decision making. The advisory group responded by trying to draw a sharp boundary between the technical issues they would explore and other concerns that the commissioners needed to investigate. The committee members observed demonstrations of the three systems, not only at the FCC's hearings but also at the laboratories of the individual companies. They analyzed and compared the performances of the systems and compared these results with the performance of monochrome television. But they left it up to the FCC to decide which color system was better. According to the committee, "no recommendation for the adoption of a specific system is given since the committee believes that the decision to adopt a system must include consideration of many social and economic factors not properly the concern of the technical analyst." The committee merely wanted to "provide a comprehensive and understandable basis on which the technical factors may be considered in arriving at a decision."[57]

The advisory committee used nine performance characteristics to judge the value of the different color systems. The members did not rank the different categories in terms of importance, but left the issue for policy makers who would need to take "into account the economic, political, and sociological factors, as well as the technical factors involved." The performance characteristics evaluated by the committee were (1) "adaptability," or how easily and effectively individuals can modify existing monochrome receivers to pick up color transmissions in black and white; (2) "compatibility," or how well existing sets reproduce an image without any modifications; (3) "convertibility," or how easily and effectively individuals can modify existing monochrome receivers to receive color broadcasts in color; (4) continuity of motion; (5) color fidelity; (6) effectiveness of channel utilization; (7) geometric resolution; (8) flicker-brightness relationship; and (9) superposition performance.

The committee decided that RCA's system clearly performed better than CBS's in three categories—compatibility, adaptability, and effectiveness of channel utilization. CBS clearly outperformed RCA in two categories—color fidelity and convertibility. Color Television, Inc.'s system did not outperform the other two systems in any of the categories, but it was judged comparable to RCA's in terms of compatibility. CBS's and RCA's systems were both ranked first in the category evaluating continuity of motion. Comparing the systems in the last three categories was more difficult. Each category included three or four subcategories and evaluations of superior performance depended on which subcategory was used. Although the committee had avoided ranking the categories, it did rank the different criteria within each category. The committee decided that, among the subcategories, large-area flicker was more important than other types of flicker; that the number of picture elements per color picture was more important than vertical or horizontal resolution; and that for superposition, registration was more important than color breakup or color fringing. Given these decisions, RCA (along with Color Television, Inc.) outperformed CBS in two more categories—flicker and resolution; CBS was judged superior to the other two systems in the last category—superposition.[58]

Like some of the earlier technical studies by other advisory committees, this congressionally inspired study helped clarify and rationalize many of the salient issues, but it also underscored the complexities involved in trying to analyze television performance objectively. The committee acknowledged that evaluations of performance depended on "the subjective reactions of the committee members to the demonstrations." Unlike other studies, this inquiry did not advocate the need for actual home tests using statistically significant samples of different consumers. The committee based its evaluation on the observations of eight individuals, under laboratory conditions. The observers judged performance using qualitative terms such as good, satisfactory, fair, poor, and excellent. The final report acknowledged that these terms were not exact and rigorous but were "in the last analysis merely words on which the committee was able to agree as being most indicative of relative performance."[59]

Unlike the Joint Technical Advisory Committee, Condon's advisory committee did not attempt to evaluate its observations in terms of levels of certainty (i.e., rating data into three classes), but it did deal with another aspect of uncertainty. Recognizing that industry experts were constantly developing television systems, it attempted to take into account the uncertainties involved in trying to predict future innovations. The development of the tricolor picture tube, which RCA demonstrated to the commission in April 1950, illustrated the difficulties involved in predicting technological progress. After it became clear that the tricolor tube could be developed, Donald Fink admitted that, only a short time earlier, he "would have said on a stack of Bibles that the tricolor tube was impossible." Condon's committee attempted to predict the "prospect for future improvement" for each of the three color systems the commission was considering. It believed "the net long-term good to the public is . . . greatest in that system which can be expected to reach the highest pitch of performance during the next few years." CBS's system, the members agreed, seemed to have "progressed furthest toward full realization of its potentialities within the confines" of the scanning standards. The committee acknowledged that the introduction of the tricolor tube meant that CBS's system could also function as all-electronic. The other two systems had "considerable room for improvement within the confines of the scanning standards"; however, the system presented by Color Television, Inc., according to Condon's committee, also had inherent limitations, including interline flicker, which experts could not easily solve.[60]

CBS Color Authorized, 1950

After evaluating all the evidence, including the report from Condon's committee, the FCC issued its own report on color television on September 1, 1950. Although three commissioners dissented with minor aspects of the report, "First Report of Commission (Color Television Issues)," all the commissioners agreed that CBS's system outperformed the other two competitors. The FCC's report did not immediately authorize commercial operations based on CBS's standards. It tried to compromise with the industry demand for a compatible system by tentatively authorizing flexible standards. This proposal, which the commission had first introduced during the hearings, would allow new receivers to pick up both established monochrome broadcasts as well as color broadcasts based on CBS's field-sequence system. Television sets operating with "bracket standards" would receive broadcasts having a variable number of lines and a variable number of fields. A switch, manual or automatic, would "select instantaneously" between the two different systems.

The FCC report stipulated that the commission would not officially authorize "bracket standards" until all interested parties had an opportunity to submit comments; the commission was particularly interested in finding out if manufacturers would go along with this proposal and build new sets incorporating flexibility. If a "sufficient number of manufacturers" agreed to the proposal, then the commission would postpone a decision about authorizing a particular color system until after new hearings evaluating improvements in the CBS system as well as any new proposals for alternative systems. The commission felt it could afford to take more time to decide about color if the industry accepted bracket standards because it would be "confident in the knowledge that adequate provision has been made to prevent aggravation of the compatibility question." However, if the manufacturers refused to accept "bracket standards," the commission warned, then a final decision adopting CBS color would immediately be issued in order to minimize the number of monochrome sets that citizens would have to convert to receive color broadcasts.[61]

In deciding that CBS's system was superior to the other two competitors, the commission relied on some of the same performance categories used by Condon's committee. As "minimum criteria" for a color system to be eligible for adoption, the commission stipulated that it should not be "marred by such defects as misregistration, line crawl, jitter or unduly prominent dot or line structure"; it should also have a "high quality of color fidelity," "adequate apparent definition," "good picture texture," and sufficient brightness for normal home viewing without "objectionable flicker." Besides these minimum performance criteria, the commission also required a system to be simple to operate for the average home viewer and "cheap enough in price so as to be economically available to the great mass of the American purchasing public." The FCC did not want to see television broadcasting support a class system with color television only "available to those who can afford to pay luxury prices." The system also needed to be inexpensive and easy to use for broadcasters and broadcast engineers.[62]

Notably missing from the commission's list of performance criteria was the issue of compatibility and the closely related categories of adaptability and convertibility. The commission argued that, "based upon a study of the history of color development over the past ten years, . . . from a technical point of view compatibility, as represented by all color television systems which have been demonstrated to date, is too high a price to put on color." The commission said it would have liked to adopt a "satisfactory" compatible color system if one had existed, but the demonstrated systems were too complex and did not produce high-quality pictures. The commission also judged that "it would not be in the public interest to deprive forty million American families of color television in order to spare the owners of seven million sets the expense required for adaptation." Some of the strongest arguments for compatibility in 1950 came from RCA engineers, but Commissioner Jones pointed out that during the 1947 color hearings, at least one RCA engineer, Elmer Engstrom, had argued that "compatibility is secondary to adequacy of performance."[63]

A second important issue not included in the commission's list of criteria for judging color was that of patents—the positions in this matter of the different companies proposing color systems. As we have seen, Senator Johnson and a number of witnesses at the FCC hearings had expressed concerns that if RCA's color system was adopted, the company would gain an insurmountable market advantage in the broadcast industry. Although the commission had investigated the patent structure of the industry during the hearings, the commission's "First Report" argued that "the decision as to whether the RCA system should or should not be adopted is based solely on a consideration of the system on the merits." Wanting to gain support for bracket standards after having alienated key elements of the industry, the FCC needed to justify its decision against RCA by emphasizing technical-performance criteria. Although evaluating these factors was by no means an unproblematic exercise involving narrowly defined technical issues, the decision to consider only "the merits" of the systems allowed the commission to avoid the kind of intense controversy that would have resulted from a decision that seemed to take into account populist concerns about monopoly control. Despite earlier statements by the commission that it believed it had clear legal authority to consider issues involving monopoly control, the "First Report" stated that "if the commission should find that a monopolistic situation does exist or such a situation should develop, appropriate proceedings can be instituted under the anti-trust laws or the commission can seek from Congress legislation to prevent the building of monopolistic patent structure in the radio field, or both."[64]

Using its list of minimum criteria as a basis for evaluating the different color systems, the commission argued that the system presented by Color Television, Inc., was unsatisfactory because it was "unduly complex" and it produced a poor-quality picture that suffered from line crawl. Like Condon's committee, the commission tried to predict whether engineers would likely solve technical problems in the future. In the case of problems in the system developed by Color Television, Inc., it decided that "improvements in apparatus will in all probability not eliminate these defects since they appear to be inherent" in the company's line-sequential system.[65]

RCA's system performed much better than Color Television, Inc.'s, but it still fell short of the FCC's minimum criteria. The commission complained that RCA's system was too complex and expensive, did not produce satisfactory color fidelity, and was marred by problems with image registration. The system was based on new principles such as mixed highs that needed further development, and the commission was not convinced that the new techniques could ever be perfected. "There appears to be no reasonable prospect that these difficulties in the RCA system can be overcome, because of misregistration, mixed highs, cross talk between picture elements, and criticalness of color control." The tricolor tube that RCA demonstrated toward the end of the hearings would theoretically eliminate registration and other problems, but it still needed to be drastically improved. The commission felt there was no assurance that the tube would enhance the performance of RCA's system and solve the problem of complex and expensive receivers. Perhaps most important, RCA's system had not undergone extensive field testing. The commission wanted to be consistent with its requirements for field testing announced in March 1947, when it had refused to authorize the former CBS color system, "a much simpler system and one which had more field testing than the RCA system."[66]

The commission emphasized field testing and the use of nonexpert evaluation partly because it did not entirely trust the testimony of industry engineers. Unlike in 1947, when a different set of commissioners evaluated color television and avoided exploring why industry engineers seemed to be inconsistent in their testimony or whether their testimony might be motivated by their employers' economic interests, the 1950 commission took a more critical perspective. Perhaps most significant, key commissioners, especially Jones, argued that RCA engineers and their supporters could not be trusted because of the inaccurate, and suspiciously self-serving, predictions they had made about their system in 1947. Jones pointed out that engineers' predictions in 1950 for RCA's newest system—that the problems noted by the FCC would soon be solved—were very similar to the optimistic predictions some of the same engineers had made in 1947 for RCA's earlier "simultaneous" system. Despite the engineers' promises that they could easily overcome registration and other problems, and that the system would be ready for full field testing in twelve months, RCA had been forced to abandon the system when it became clear that the problems were overwhelming. Jones believed that RCA's "engineering testimony in 1946–47 is rendered so completely worthless by the 1949–50 record that the kindest that can be said in explanation is that their economic interest blinded their engineering judgment." Although the official commission report was more circumspect, it did indicate that its final decision against RCA color was influenced by the past testimony of RCA engineers. "In weighing those recommendations and expert opinions," the commission reported, "we cannot overlook the fact that many of these same parties offered recommendations and expert opinions of the same kind as the basis of their advocacy in the 1946–47 hearing of the simultaneous system—a system which never survived field testing."[67]

Especially in comparison with the other two systems, CBS's field-sequence system, according to the commission, "produces a color picture that is most satisfactory from the point of view of texture, color fidelity and contrast." It was also simple to operate and appropriate for home viewing, and the commission was impressed with the field testing the company had conducted under "widely diversified circumstances." Although the commission acknowledged that CBS's system was not perfect, it emphasized that the system's problems were not very serious. It recognized that, compared with the existing monochrome system, CBS's system was more susceptible to flicker as the brightness was increased, but the commission argued that the system was able to produce sufficient brightness levels for home use without causing flicker. It also thought enough evidence existed that engineers were developing new, long-persistence phosphors that would allow for the use of even higher brightness levels without increasing the risk of flicker. The geometric resolution of CBS's system was inferior to both monochrome television and the other two color systems, but the commission believed that, at least compared with monochrome, "the addition of color more than outweighs the loss in geometric resolution."[68]

A more serious problem, according to the commission, was that the size of tube that could be used with the color disc was restricted to less than 12.5 inches. If engineers could develop a satisfactory tricolor tube for CBS's system, this size limitation might be overcome. But since the tricolor tube was still in the nascent stages of development and had never been demonstrated with CBS color, the commission could only speculate about how it might perform. This uncertainty about the use of the tricolor tube with CBS color was an important reason the commission decided to give the industry an opportunity to delay a final decision about CBS's system if it agreed to authorize bracket standards. In minor dissents from the "First Report," Commissioners Jones and Hyde favored immediately authorizing CBS's system without investigating bracket standards, and Commissioner Hennock favored waiting three months to give the industry an opportunity to demonstrate its most recent developments. But all six commissioners were confident that even if the industry refused to go along with bracket standards, CBS color could be satisfactorily implemented. They felt optimistic that technological progress would solve any problems that still existed with CBS's system. The commission's report remembered that the industry had "succeeded in creating much larger tube sizes than those demonstrated in 1941 when standards for black and white television were adopted."[69]

Archival records indicate that the commission made its 1950 color decision based on recommendations from its engineering staff. Unlike the 1947 decision, the commissioners and staff generally agreed about the importance of a timely color decision, even if it meant overturning the status quo. The chief engineer wrote in a memorandum in late June 1950 that CBS's system was "the only system that we can safely adopt from a technical point of view today." But he also acknowledged that because all three systems were still "subject to considerable further technical development," "the ideal solution" would be to wait "another year or so for further technical developments on all three systems." The compromise solution of flexible "bracket standards," proposed in the commission's report, was based on a recommendation from the chief engineer.[70]

The evaluations of different performance characteristics in the 1950 report also originated with the engineering staff. Like Condon's committee, the staff constructed a table comparing the performance features of each system. The FCC engineers used many of the same categories in their evaluations, but they also considered the cost and complexity of the color systems. While Condon's committee avoided judging the relative value of different performance characteristics, the engineering staff assigned a numerical value indicating the importance of each category to the overall evaluation. They also rated on a similar scale the performance of the different systems in each of the categories. CBS received the highest possible score for six of the categories, including color fidelity, receiver cost and complexity, and picture quality and texture. The engineers also assigned the highest level of importance to the first two of these categories. RCA did not receive the top evaluation for any of the categories. Color Television, Inc., performed poorly in nearly all the categories.[71]

Although the commissioners had criticized industry engineers serving on the system committee for trying to usurp the commission's role by making policy recommendations about such issues as the importance of compatibility, they did not criticize their own engineers for acting in a similar manner. The engineering staff played an important role in the commission's decision to deemphasize compatibility when evaluating the color systems. The chief engineer argued that compatibility was an issue "that deserved short term consideration and not long term consideration." Since he believed it was the responsibility of his "office and the immediate divisions connected with it such as the Laboratory, Frequency Allocation and Treaty, and the Technical Research Divisions to look at the long-term trend rather than the short term," he recommended that compatibility should not be considered an important criterion. The motivation of the FCC's engineers for taking an active role in the color hearings and final decision is clear from a staff memorandum in which one engineer advised that "insofar as possible do not put our engineering problems up to them [the commissioners] to decide. . . . The commissioners cannot remember all the details of engineering of all the services. . . . We are the experts in television and should know the service thoroughly." This statement supports observations other scholars have made about the important role of the staff, especially engineers, in formulating and writing commission opinions. The activist role of the FCC engineers and the general support for color television among the commissioners led the chief of the Laboratory Division to take the initiative and construct an adapter that could easily and cheaply be used to facilitate reception of broadcasts from both CBS color and existing monochrome stations.[72]

The Laboratory Division was part of the Office of Chief Engineer. Two other major branches of this staff office (the Technical Research Division and the Frequency, Allocation, and Treaty Division), along with the Office of General Counsel, provided the commission with "top level professional assistance." The Office of General Counsel advised the commission on all legal issues, including interpretations of the Communications Act. In addition to these two high-level staff offices, the commission operated three "operating bureaus" organized to deal with the major industry groups: a Broadcast Bureau, responsible for radio and television broadcasting; a Common Carrier Bureau, responsible for the telegraph and telephone industry; and a Safety and Special Radio Services Bureau, responsible for nonbroadcast users of the radio spectrum. A separate bureau, the Field Engineering and Monitoring Bureau, conducted field investigations, monitored the radio spectrum, and examined radio operators.[73]

In actively participating in the color decision, the commission's engineers tried to deal with the complexities and interrelationships of the issues in a judicious and sophisticated manner. They acknowledged that the technical issues were by no means straightforward or clear-cut. The chief of the Technical Research Division pointed out to the chief engineer that "the technical populace is far from unanimous in the estimate of which constitutes the best bet." He emphasized that they needed to take into account the different objectives and try to understand the "conflict and interplay" among the different possibilities.[74]

Unlike the engineers guided by a technocratic perspective during the 1920s, the engineering staff in 1950 was particularly interested in trying to take into account the broad implications and the "long term effects" of technical decisions. They took an expansive view of engineering and their role as engineers. The division chief quoted above also emphasized that the commission should clearly detail the commission's objectives and how the objectives would be affected by its color decision. He stressed that the commission should "go into considerable detail on the risks avoided and the risks taken." Had the commission done a better job in the past thoroughly discussing objectives and calculated risks, he believed, it could have avoided recent controversies such as the debate about sky-wave and tropospheric propagation and issues leading to the television freeze. The staff member used his own version of risk analysis to support the argument that CBS had the best system. "The risk that this system can be much improved appears to be a greater risk than that the presently proposed compatible systems can be made to be superior," he wrote, "but in this former case one might take the risk and lose and still be left with a usable system. This would not be true of the latter risk."[75]

The Commercial Failure of CBS Color

After the FCC announced its decision on September 5, it gave the industry until the end of the month to agree to build bracket standards. The commissioners assumed the industry would view this alternative as more desirable than having CBS's system immediately declared the national standard. But the major manufacturers believed there was another option, and they refused to support either bracket standards or CBS color. Despite having indicated during the hearings that they would go along with the FCC's final decision, the manufacturers decided during a special meeting of the Radio Manufacturers Association in the middle of September to support the system committee's efforts to produce a better color system before CBS's system had an opportunity to gain acceptance. They officially argued that even if engineers could build television receivers with bracket standards, they would be too expensive and complex; unofficially, they felt they had not been treated fairly by the commission during the hearings, and they feared that an unfortunate precedent would be established if they allowed the FCC to usurp the industry's standards-setting role.[76]

After the FCC's deadline passed and the industry's intentions became clear, the commissioners voted three to two in favor of following through on its ultimatum and declaring CBS color the national standard. But the industry also carried out its threats by actively trying to thwart CBS's efforts. In October, during another meeting of the manufacturers association—now known as the Radio and Television Manufacturers Association—RCA announced that it would immediately file suit against the FCC to overturn the color decision. It also used this meeting to consolidate its position by receiving assurances from the networks—NBC, DuMont, and ABC—as well as many of the other manufacturers that they would not support CBS. CBS had been denied permission to attend this special meeting. The FCC could not stop the industry's efforts to block CBS color because it had not required the industry to phase out monochrome operations and adopt the new system; the decision had only given CBS or any other interested company permission to begin commercial color broadcasting using CBS's standard.[77]

RCA's lawsuit challenged the FCC's decision based on four complaints. The company argued that incompatibility violated the 1934 Communications Act because it was not in the public interest. It charged that E. W. Chapin, the FCC engineer who had taken the initiative and built an adapter for CBS color, had interpreted technical information in a way that supported his former employer, CBS. RCA also contended that by refusing to use the NTSC, the commission had violated the Administrative Procedures Act, which dictated that government agencies had to be consistent in formulating policy. Since the commission had used the system committee to help determine monochrome standards, according to RCA, it should have used it as well for color. Finally, the company charged that the FCC did not in fact have legal authority to regulate manufacturers. The suit eventually made its way to the U.S. Supreme Court, which decided in May 1951, in a seven-to-two opinion, in favor of the commission. The court rejected all four of RCA's complaints. The ruling not only established clear legal authority for the FCC to set manufacturing standards but emphasized that the FCC had a legal duty to take this kind of action when necessary. The Court did not think the commission had violated the public interest or the Administrative Procedures Act and it did not think there was evidence of blatant bias. The court pointed out that a number of engineers on the commission were also former employees of RCA.[78]

Although RCA lost the legal struggle against CBS color, it won a victory in the marketplace. There it had an advantage. A lower court had ruled in December that CBS should wait until the RCA lawsuit was resolved before beginning commercial operations. During the period when the courts considered the suit, monochrome sales continued: the number of television receivers owned by the public increased by 50 percent. When CBS finally began commercial operations in June, it had to compete with an industry more entrenched than ever.[79]

Although the first commercial broadcasts during summer 1951 from its station in New York City attracted considerable attention and support, the company increasingly had difficulty attracting sponsors for the expensive programming. In the early days of monochrome operations, when the audience was small and sponsors were difficult to find, RCA had sufficient economic resources to support television broadcasting. CBS was at a disadvantage because it was not a large, diversified company. It wanted to continue broadcasting its monochrome programming, which was finally beginning to make a profit, and since it was not possible to transmit together separate programming with incompatible standards, CBS essentially had to operate simultaneously two networks, one in monochrome and another in color. Also, unlike RCA, CBS did not have manufacturing capabilities. Because manufacturers mostly refused to support its color operations, the company decided to acquire, in April 1951, a small company, Hytron Electronics, that had recently begun to manufacture television receivers after having produced radio tubes for a number of years. But the conversion to color manufacturing proved to be more expensive and complex than first realized.[80]

An especially important factor hampering CBS's efforts to establish color broadcasting was the country's involvement in the Korean War. Manufacturers who had traditionally supported CBS, such as Zenith and Westinghouse, were hesitant to get heavily involved in a new endeavor because of uncertainty about the effect of the war on domestic manufacturing. CBS's efforts to introduce color television on a large scale essentially ended in the fall of 1951 when the Office of Defense Mobilization placed a ban on the production of color-television equipment.

Some evidence exists suggesting that CBS was not entirely disappointed when the ban was placed into effect; the company may have seen the ban as a good excuse to abandon a failing endeavor honorably.[81]

CBS's troubles led RCA and the remainder of the industry to increase their efforts to come up with a compatible color system that the FCC would be willing to authorize. During 1952, the system committee used new techniques developed by the industry to improve RCA's dot-sequential system. By July 1953, the commission was sufficiently impressed with these efforts to reconsider its previous color decision and evaluate the new system. When CBS announced that it would follow most of the rest of the industry and support the color system revised by the system committee, the commission quickly acted to overturn its earlier ruling and authorize this compatible system. The FCC announced its final order setting color standards on December 17, 1953. Color television now had the support of the major manufacturers and broadcasters and was ready to take off.[82]

But like FM radio (at least, FM as it was developed in the United States), color television did not catch on with the public for at least another decade. RCA and other companies had helped delay color television by arguing that CBS's system was inherently inferior and needed further development; ironically, they now found that their own new system also needed further development. Some of the same problems that the FCC had identified in 1950 with RCA's system remained in 1953, including difficulties with image registration and cost. Although CBS's field-sequence system never succeeded in a marketplace dominated by RCA and other manufacturers and broadcasters committed to compatibility, it continued to be used in special circumstances, for example, at medical centers and the U.S. Space Program. The National Aeronautics and Space Administration chose an improved version of CBS's system to use with the Apollo Program in the 1960s and 1970s.

Without falling into the fallacy of counterfactual history, we can conclude that CBS's system might have been successful had it received more support from the industry and the FCC. It probably stood the best chance of success in 1947, before monochrome television became completely entrenched. If the FCC had authorized CBS color in 1947 and if CBS had done a better job gaining support from manufacturers well before this date, CBS likely could have adapted its system to achieve commercial success. Testimony at the various hearings had underscored the difficulties of arriving at objective criteria that would reveal the one best system. Consumers' expectations of picture quality had changed over time with new developments. CBS's color demonstrations were at least as good as the earliest monochrome broadcasts, which had been accepted by the public.

As in the case of monochrome television, the development of color television was shaped by a variety of institutions and individuals, including manufacturers, broadcasters, and government officials (especially officials from the FCC) carrying out a mandate to establish commercial standards. In both cases, Congress also played an important role by encouraging the FCC and the industry to resolve differences and achieve closure. Participants needed to finesse the tension between, on the one hand, trying to find the right time to set standards to avoid freezing them at an inferior level and, on the other hand, satisfying the immediate demands of a consumer society for new products. In the case of color television, this also involved evaluating the importance of having a system compatible with existing monochrome. In both cases, we see the significance of contested meanings of *technical* evaluation and different views about the role of experts in policy making. For both technologies, key commissioners encouraged participants to deal with the complexities of technical evaluation by taking into account the socioeconomic impact of the new developments. Although this resulted in the FCC examining issues involving patent and monopoly control, the final decisions played down the importance of these issues by emphasizing that they based their final judgment on an evaluation of the technical merits of the different systems.

There were, however, also major differences in how standards were established for the two technologies. Unlike color television, monochrome did not have to compete with an established commercial system. In this respect the history of color television more closely resembled FM's development. Also, monochrome television was mainly pushed by a company (RCA) that had both manufacturing and broadcasting capabilities; the main advocate for color (CBS) was primarily a broadcaster that did not have the extensive resources of a large diversified company. Although in both cases, the FCC did not entirely trust the judgments of industry engineers as technical advisers, the 1950 commission was far more critical. As a result, FCC engineers played a more important role in deciding the color decision. The involvement of industry engineers was minimized. The FCC had encouraged industry engineers to come together to reach consensus in support of the monochrome decision. With the color decision, the commission did not think the industry experts would be able to reach a satisfactory agreement; they responded by rejecting the offers of assistance by the NTSC. With color, the commission also placed a stronger emphasis on field testing and the use of nonexpert evaluation. Further, unlike the monochrome decision, the color decision was complicated by other factors that had to be evaluated at the same time, including the status of UHF, educational television, and the lifting of the freeze on new stations. Even more than the monochrome decision, the color decision illustrates the difficulties and tensions involved in trying to evaluate technical uncertainty and predict technological development.

)))

Epilogue

It should be clear that the electronic-media industry—that influential and powerful force in modern life—was not necessarily an inevitable result of technological progress or economic development divorced from political and social context: key individuals and institutions made decisions that helped shape the industry. The focus on the establishment of standards for radio and television in the United States and the connection to public policy underscores the need to understand complex negotiations among different groups and individuals in the history of technology. Some of the essential developments in broadcasting—including commercial support, network structure, and limited government oversight in the public interest—are still with us today. This epilogue pulls together major conclusions of this book and ties the historical discussion to recent and related events in the history of U.S. broadcasting and broadcast policy.

Although commercial broadcasting in the United States is based on free-enterprise principles, from the beginning the system has depended on active government involvement. A traditional image of polarized choices between private and public realms does not fit the historical realities of broadcasting in the United States. As this study underscores in particular, an important role has been played by the new organizational sectors developed, especially by Hoover, during the 1920s. Policy makers assumed that the introduction of radio and television primarily involved issues of technological innovation and scientific evaluation. Technical experts from both government and industry came together to standardize and bring order to the broadcast industry using technocratic principles. Technical issues such as propagation conditions, spectrum scarcity, and signal interference tended to circumscribe traditional policy and legal concerns relating to monopoly control and political authority. Experts thus played an important role finessing dilemmas inherent to corporate liberalism, including the tension between individualism and corporate collectivism as well as the dialectic between neutral legalism and pluralist decision making.[1]

The discussion has focused on how participants in policy making managed these and related dilemmas when they oversaw the introduction of new innovations in broadcasting. The central problem of using technical experts was itself problematic and was handled differently throughout this period. Decision makers wanted to use experts for different, yet related, reasons. On the one hand, they placed much weight on the advice of engineers and scientists because they were the ones best equipped to make sense of complex policy decisions that demanded knowledge of physical facts and principles. On the other hand, they also sometimes contended that these same individuals were best equipped to solve all policy issues, even when they clearly also involved social and economic considerations, because they believed the professional experts were more likely than others to be unbiased and objective.

A minority of individuals expressed views representing a countertradition to this general faith in expertise; these views pointed to additional complexities that policy makers needed to consider. We saw in chapter 1, for example, how an important senator argued during the earliest debates about establishing an independent regulatory commission that engineers would not make good candidates for the proposed commission because they would not be able to lift themselves above technicalities and deal with large fundamental issues. He believed their narrow training and experience would preclude such actions. Other individuals contended that policy decisions should be based on the experiences of the average radio listener or television viewer. They wanted decision makers to evaluate technical standards by undertaking field tests using consumers operating equipment under home conditions, instead of relying on scientific evaluation by experts in laboratories. Amateurs or hobbyists also believed they could do a better job than professional experts evaluating new sets of standards for equipment. But the commission, when it was formed, did not heed the wishes of many of these groups or individuals. The amateurs, in particular, found themselves increasingly marginalized throughout the period covered by this study. Key individuals also pointed out that consumers operating in the marketplace would not necessarily provide democratic evaluation; the voices of wealthy consumers would count more than others because they would be the only ones able to purchase the expensive new equipment. And even if extensive field tests were attempted prior to commercial marketing of home receivers, experts would still be needed to set up the tests and analyze the results scientifically.

The problems involved in using experts also underscores important populist themes in the history of broadcasting. Policy makers had to consider populist anxieties about the influence of big business and the development of centralized control throughout the early history of radio and television. Concerns about monopoly power involved both economic and political issues. The major focus of antimonopolist concern throughout the entire period of this study was the Radio Corporation of America. Individual citizens, rival companies, and government officials not only worried about illegal economic consolidation but also about the possibility that one company might gain control over the dissemination of information to the public. But populist anxieties were problematic and could in fact be used to support efforts leading toward economic consolidation and centralized control. Government and industry officials argued that clear-channel, superpower radio stations interconnected through national networks were necessary to provide service to rural listeners, a traditional source for populist views. They tended to ignore the fact that local, noncommercial radio stations operated by educational institutions had already been providing service to many rural listeners. Because the legal authority for regulatory commissions to take into account issues of monopoly and patent control was open to debate, these issues tended to be circumscribed by questions of neutral technological requirements related to the development of the best set of engineering standards.

A kindred dilemma that policy makers needed to engage was the tension between advocacy and objectivity. Although the federal government became more important after the war as a source of research relevant to broadcasting, much of the technical information needed to evaluate broadcast policy continued to be held by the corporations that employed many of the country's radio and television engineers. These corporations were in many cases using technological innovation as part of a business strategy to gain an economic advantage by acquiring patents for new systems. Engineers obviously played a key role in these activities. As they moved into management positions overseeing research organizations, the experts were increasingly under pressure to demonstrate loyalty to their companies' policies. Government officials had good reason to doubt engineers' claims that they could set aside their employers' interests and make decisions independently, as members of professional associations.

The issues of "objectivity" and "bias" were viewed in different ways. Some participants saw these as moral problems, involving honesty and professional integrity. Engineers pointed to their professional training and commitment to scientific truth to justify a belief that they would act with more honesty than other individuals. Skeptical officials may not have thought that the experts were biased in the crudest sense of the term, but they did think arrangements should be made to manage the dilemma. A chairman of the Federal Communications Commission, James Lawrence Fly, established the National Television System Committee as an attempt to handle the problem through a new institutional arrangement. Other participants, especially from smaller companies seeking to gain a market advantage, did not accuse rival engineers who made judgments that benefited their employers of being dishonest, but they did raise the possibility that they could be guilty of self-delusion or of not having a capacity for self-criticism. Officials from Philco, for example, wanted government officials to make sure all relevant information, including the patent positions of different companies, was made public so that decision makers could make a fully informed judgment about black-and-white television standards. Engineers might be influenced by their employers' interests without being completely aware of that influence.

Especially beginning in the 1930s with Commissioner Fly and other New Deal officials, policy makers did become more critical and self-reflective about issues connected to the tension between objectivity and advocacy. This trend continued into the postwar period of television planning—although government officials during the 1940s and 1950s were less likely to take a stand against the dominant interests in the industry, with some notable exceptions. Engineers advising policy makers also became more cautious and self-critical during the postwar period. They tried to acknowledge, explicitly, the tentativeness of their judgments and attempted to take into account the uncertainties involved in trying to predict technological developments. However, although these engineers were more willing to admit that technical considerations could not easily be separated from social factors, they still tended to believe that they were best suited as truth-seeking professionals to evaluate these hybrid considerations. This was particularly true for engineers with the FCC; their influence, especially with the 1950 color decision favoring CBS, reflected the growing importance of engineering work done by the government. In the 1950 decision, they tried to evaluate possible risks involved with different policies in order to find an ideal balance between different options. In seeking a more sophisticated view of the complexities of policy making for broadcasting, they also tried to take into account different levels of uncertainty of technical knowledge.

Despite these developments, technocratic values still played a major role in policy making for radio and television broadcasting throughout most of the period analyzed in this study. We see this especially in the way officials tried to legitimate complex decisions in terms of narrow technical concerns. Boundaries were constructed between technical and nontechnical criteria to facilitate policy-making processes and give authority to final decisions.

This book supports the work of other scholars who argue that the technological enterprise should simultaneously be viewed as social, economic, political, and organizational. Like other engineers, the technical experts in this book practiced "heterogeneous engineering, the engineering of the social as the physical world." Engineers and inventors have "so thoroughly mixed matters commonly labeled economic, technical, and scientific" that their "thoughts composed a seamless web." The systems concept used by Thomas Hughes and other historians helps make sense of the interrelationships between categories. Entrepreneurs and system builders studied in this book helped construct the U.S. system of radio and television broadcasting that depended on both free enterprise and active government involvement.[2]

The use of Hughes's term *seamless web* helps make sense of the tension between technocratic and nontechnocratic views analyzed in this book. Although the technological enterprise is simultaneously "social," in the broadest sense of this term, engineers and other officials involved in the creation of systems need actively to maintain boundaries between these two realms to maintain stability, authority, and independence. This "boundary work"—a useful phrase developed by historians and sociologists—is an essential aspect of "heterogeneous engineering." Boundary work is a crucial activity that complements such efforts as the creation of a need for particular technological developments. Experts and public officials worked to maintain boundaries in order to achieve closure and avoid public controversy that would play up contingencies and uncertainties. I do not mean to imply crass duplicity on the part of engineers and other officials. Boundary work is a subtle activity, not necessarily directly the result of clear reflection and planning. Individuals were probably for the most part unaware that by applying technocratic principles to policy making about radio and television standards, especially by attempting to reduce issues to narrow technical facts, they were indirectly supporting corporate liberal principles.

In many cases, boundary work needs to be understood in conjunction with a lack of a critical understanding by engineers and other participants of the complexities of the engineering enterprise and its relationship to social developments. The perception among radio engineers that their profession was more scientific than other groups, which had its origins in the founding of the Institute of Radio Engineers, played an important role in processes connected to boundary work. Radio engineers not only attempted to draw a sharp boundary between technical and nontechnical considerations or facts and values in order to maintain authority and autonomy, but also undertook to develop standards using negotiations that often implicitly recognized the hybrid nature of this activity. Apparent inconsistencies in public statements by these engineers reflected this complexity.

An emphasis on how certain technologies are inherently superior and the rhetoric of technological utopianism expressed by policy makers, engineers, and other observers predicting the role of innovations in broadcasting also reflected the importance of technocratic values as well as the need to engineer the social aspects of the technological enterprise. The utopian predictions assumed that technological progress in broadcasting would essentially define general cultural and social progress. Instead of seeing radio and television as means toward clearly defined ends for society, many individuals viewed the new innovations as ends in themselves; the introduction of new broadcast technologies would be sufficient to solve major problems, ranging from world conflict to the threat of monopoly control. Ironically, the new technologies provided an ideal medium for promoting these utopian themes, which also served as a powerful marketing program for companies seeking to use technological innovation as a business tactic.

The evolution of new communications media to tie individuals and communities together has been closely linked to social changes that have brought greater complexity and large-scale development. Radio and television broadcasting has helped individuals adjust to new cultural patterns connected with the rootlessness of modernity. As the constant changes of modernity have eaten away at the local connections of traditional communities, broadcasting has helped fill a void with the development of a shared mass culture. Like consumerism, which it helped support, especially through the common medium of advertising, broadcasting has helped allay the anxieties and disruptions of modern life. But the utopian predictions that have been a common theme in the history of broadcasting appear problematic in the context of actual historical developments. Instead of helping to build community, for example, broadcasting has in many cases reinforced individualism: consumers have found meaning in the private and individual experiences of watching television or listening to the car radio.

New technological developments in broadcasting since the 1950s have presented challenges to old regulatory patterns in the United States; they have also reflected new views of the role of government regulation in communications. Cable television, satellite communications, and digital technology are among the most important new developments in broadcasting. Beginning especially during the 1970s, a philosophy of deregulation has also become dominant. Technological and political changes helped inspire Congress to rewrite the 1934 Communications Act. The new 1996 Telecommunications Act aims to unleash market forces in order to stimulate new technological developments leading to the convergence of broadcasting, cable, telephone, and computer technology.

Cable television dates back to at least 1950, when local systems began to use large antennas to transmit broadcast signals to homes unable to pick up nearby stations. These "community antenna television" systems charged customers a fee to amplify and transmit television signals over coaxial cables. Engineers at Bell Laboratories had perfected coaxial cable during the 1930s. Networks used the new technology, which could transmit a multiplicity of signals within a single wire, to interconnect affiliated stations.

During the 1950s, broadcasters did not view cable television as a threat, even as operators began to use microwave relays to import signals from distant stations for use in regions that had only limited television service. Despite the efforts of the FCC to establish a diverse national service for the entire country, as late as 1958, 34 percent of homes could receive only one television signal. At first cable companies used microwave relays to supplement locally available television service. But during the early 1960s, operators increasingly began to import signals from distant, large cities. Local broadcasters, in particular, saw this action as a threat because it increased the competition they had to face. They pressured the FCC to intervene in order to protect "free," over-the-air television, especially in smaller markets.

Commissioners responded positively to the broadcasters' requests for support during the 1960s. UHF stations in small cities were especially vulnerable to competition from cable. During this period, both the commission and Congress were searching for ways to save UHF stations from their precarious existence as second-rate broadcasters. In 1962, Congress passed legislation forcing manufacturers to include both UHF and VHF controls on all television receivers. The commission also developed regulations during the mid 1960s to protect local broadcasters from cable television. A 1965 rule required cable operators to include all local stations in their transmissions to subscribers. Another regulation limited the ability of cable systems to import distant stations. After a cable-television company challenged the FCC's jurisdiction, the U.S. Supreme Court ruled, in 1968, that the commission did have authority to regulate the industry. In 1972, the FCC issued "definitive cable regulations" that further required cable systems to provide at least three local channels, one for use by local governments, a second for community access, and a third for general education.

However, new court rulings and changing views about regulation during the 1970s eventually helped undermine support for the regulation of cable television. In 1980, for example, the commission did away with rules limiting the importation of signals from distant cities. Congress gave the FCC clear legal jurisdiction over cable in 1984 when it passed the Cable Communications Policy Act. The new act generally affirmed new deregulatory policies already in place.[3]

Satellites, like cable, served as new delivery systems for broadcasting. Satellites were especially important in supporting the growth of cable as a separate service able to compete directly with broadcasting. The use of satellites for broadcasting television programs to other countries had been a major aspect of one of the first proposals for artificial satellites, published by Arthur C. Clarke in 1945. Communications policy became closely linked to space policy after the launch of the Soviet Union's Sputnik in 1957 and the establishment of the National Aeronautics and Space Administration (NASA) in 1958. NASA and other agencies of the federal government, particularly the FCC, recognized the importance of communications satellites and debated national communications-satellite policy. A major issue for debate involved the question of private versus public ownership. The 1962 Satellite Act established the government's position on communications satellites and helped to end interagency rivalry. The act established a new private corporation, the Communications Satellite Corporation (COMSAT), with authority to set up a satellite-communications system. Although the legislation rejected government ownership, the new company was subject to regulation by the government and was mandated to take into account the public interest. In 1965, COMSAT established the International Telecommunications Satellite Organization (INTELSAT), a global communications-satellite system with international participation. As the major shareholder, COMSAT maintained executive authority over the new organization.[4]

The American Broadcasting Company (ABC) made the first request for domestic broadcast-distribution service using satellites in 1965. However, COMSAT officials initially opposed any new system, domestic or international, that would be outside their authority. They held to the belief that only one global system was appropriate. The FCC opened an inquiry on the question of whether to allow the use of satellites for domestic broadcasting in 1966, but did not make a final decision until 1972. By this date, fourteen companies had filed applications for the use of different systems. The commission's final judgment affirmed an open-skies policy that had been developed by White House policy makers during 1970. The domestic communications-satellite policy authorized the use of satellites for broadcasting and encouraged competition between different systems. To prevent the American Telephone and Telegraph Company (AT&T) from using its dominance in domestic telecommunications to gain control over satellite-distribution services, the government did not allow the company to operate its own satellite system for seven years.[5]

Despite the early interest of ABC and other television networks in satellites, new pay-cable services, especially Home Box Office (HBO), pioneered their use; the success with cable television helped convince the networks that satellites would provide delivery systems superior to cables, microwave relays, and other traditional tools used on Earth to link producers to consumers. The use of satellites by cable operators also helped support the growth of that industry. The first use of a satellite by HBO occurred in 1976 with RCA's *Satcom I.* The demand for communications satellites by the pay-cable and broadcast networks as well as a new service that broadcast directly from satellites to individual homes equipped with special receivers helped stimulate many more launchings during the 1980s. By 1988, thirteen different satellites were being used to transmit more than one hundred programs-by-cable services.[6]

The 1980s saw the development of another innovation: digital/high-definition television. Unlike new delivery systems such as cable and satellites, this represented a new form of broadcasting based on new standards. The National Association of Broadcasters promoted high-definition television in the United States during the 1980s. The association championed the new system partly to prevent land-mobile radio services from gaining control of UHF television channels. Broadcasters argued that they needed the underutilized UHF channels for the new and improved television system. After signaling that it was ready to approve land-mobile's request for UHF channels, the FCC announced in 1987 that it favored using the new channels for a new high-definition television standard.

The movement for a new standard received strong political support when influential U.S. politicians learned that the broadcasters were considering adopting a Japanese system. The first demonstration in the United States of the Japanese system had been sponsored by the National Association of Broadcasters in early 1987. U.S. television manufacturers had been unable to compete with Japanese companies during the 1970s and 1980s; now it appeared that the Japanese would win the race for this new system and the industry it would nurture. In response, Congress pressured the FCC to find ways to encourage U.S. companies to develop their own high-definition systems. The commission's decision to support high-definition television thus needs to be understood in the context of both domestic competition among users of the electromagnetic spectrum and international competition between domestic and foreign economic rivals.[7]

The commission established an advisory committee in fall 1987 to promote competition among different systems and evaluate the developments. Contestants in the race for high-definition television included not only traditional manufacturers of analog systems but also companies interested in developing a digital standard that would encode information in binary terms. Once it became clear that digital television was possible, the commission supported the development. Computer companies were especially interested in digital, high-definition television because of its capacity to interface with computers and potentially provide interactive broadcasting. In order to make the system fully compatible with computer monitors, manufacturers wanted the commission to authorize a high-definition standard that included progressive, rather than traditional, interlaced, scanning. Since television manufacturers held the patent rights to progressive scanning, the final decision needed to take into account a number of different kinds of considerations.

The commission eventually helped convince the companies involved in the race for high-definition television to form a compromise "Grand Alliance" system. The advisory committee was in charge of testing and evaluating the final system authorized by the commission. Broadcasters conducted the first regular experimental broadcasts using the new system in the summer of 1996. The commission had mandated the conversion of television to a high-definition standard five years earlier. The 1991 order gave stations a two-year period after the development of a new standard by the advisory committee before they would have to apply for a new UHF channel for high-definition broadcasting. After applying for new licenses, stations would have another three years to set up the new equipment in preparation for regular high-definition transmissions. The commission expected broadcasters to complete the process of conversion by the year 2008, when transmissions using the old system on the original channels would cease. If broadcasters and manufacturers continue to support the process, this important decision means that consumers will eventually need to buy new high-definition/digital receivers or purchase converters to add to their old sets for reception of the new broadcasts. The converters, however, will not produce true high-definition quality.

Government officials believed digital technology would help bring about the convergence of communications systems in the support of the internet, or the "national information infrastructure." During the early 1990s, the federal government promoted this new form of universally accessible and interactive communications. The 1996 Telecommunications Act reflected not only new technological developments but also the new philosophy of deregulation. Especially during the presidency of Ronald Reagan, marketplace conservatives argued that any benefits resulting from government regulation had been eclipsed by economic costs. In highly ideological terms, they argued that government intervention should be done away with in order to unleash competition. The FCC chairman from 1981 to 1987, Mark Fowler, contended that there was nothing special about television to justify government regulation. He believed it should be treated like any other consumer electronics device. In a famous speech, Fowler argued that the commission should view it as simply a "toaster with pictures." Under Fowler's chairmanship, the commission decided against continuing the tradition of setting technical standards for new innovations in broadcasting. In the case of stereo AM broadcasting and satellite direct broadcasting, commissioners decided to let "the marketplace" determine final standards for competing systems. The belief that the radio spectrum should not be treated as a private property resource has also been challenged for the first time. Although the status remains intact for broadcasters, other parts of the spectrum have been auctioned off for use by cellular phone companies and other users.[8]

Despite these new trends, traditional themes dating from the earlier period analyzed in this book are still important. As the examples of satellite communications and digital television demonstrate, the federal government has continued to play an important role in stimulating technological innovations. Despite the tendency of the FCC during the 1980s to refuse to establish technical standards, other branches of the government have become more active in international standard-setting agencies such as the International Consultative Committee for Radio. Further, although economists now play a more important role in regulatory decision making—to some extent replacing the earlier work of engineers—the theme of technocracy is still relevant. Officials tend to view economists as objective technical experts able to solve complex policy problems through scientific evaluation of economic facts. The theme of technological determinism also continues to be a major driving force. A number of officials have argued that technological change—in particular, in digital technology and the use of computer technology with the telephone or cable television—has directly caused the need for new policy and new legislation. Some individuals believe these new technological developments will solve long-standing problems and make traditional regulation irrelevant.[9]

A final issue that demonstrates the continuing relevance of some of the major historical themes analyzed in this book is the effort to manage the tension between advocacy and objectivity. Since the 1950s, officials have attempted to develop policies and legislation to avoid conflicts of interest and a "revolving door" mentality to prevent engineers and other officials from moving smoothly from government regulatory agencies to positions in the industries they previously regulated. Further, officials have drawn on the critical view of the role of technical experts in policy making that began to develop during the period analyzed in this book to support a more sophisticated view of the relationship between decision making on policy and technical evaluation. Officials are more likely to recognize the complex, hybrid nature of regulatory decision making as well as the interpenetration of facts and values—especially in the many clear cases where problems involve high levels of technical uncertainty and flexibility.[10]

A form of countertradition to the theme of technocracy also still exists and has been transformed by new developments. Especially during the 1970s, the period when deregulation was gaining influence, the commission began to allow citizens' groups, such as a grassroots organization of mothers concerned about violence and advertising aimed at children, to participate in its deliberations. Whereas earlier, officials assumed they could best represent the public interest, this new trend favored a form of direct participatory democracy. Traditional proregulation liberals who believed the commission had been "captured" by the longer-established communications industries also argued that conservative policies had acted to support the status quo by stifling innovations that might expand possibilities and enhance democratic participation. The call for deregulation has thus come from a coalition of diverse groups and individuals.[11]

)))

NOTES

Abbreviations

For reasons of space, some document titles have been shortened.

b. box; coll. collection; fldr. folder; ser. series

Armstrong Papers: Edwin H. Armstrong Papers, Columbia Univ. Archives, N.Y.C.

Clark Coll.: George Clark Radioana Collection, Archives Center, Smithsonian Institution, Washington, D.C.

Col. TV Hist./FCC, NA: Records relating to history of color television, FCC records, Record Group (RG) 173, National Archives, Washington, D.C.

Dellinger Papers: Personal papers of John H. Dellinger, records of National Institute of Standards and Technology, RG 167, National Archives, Washington, D.C.

Dellinger Rec.: General records, J. Howard Dellinger, records of National Institute of Standards and Technology, RG 167, National Archives, Washington, D.C.

Dock/FCC, NA: Docketed case files, FCC records, RG 173, National Archives, Washington, D.C.

DuMont Coll.: DuMont Collection, Archives Center, Smithsonian Institution, Washington, D.C.

ExDir/FCC, NA: General corresp., Office of Executive Director, FCC records, RG 173, National Archives, Washington, D.C.

Fly Papers: James Lawrence Fly Papers, Columbia Univ. Archives, N.Y.C.

Hoover Lib.: Herbert Hoover Presidential Library, West Branch, Iowa

Jansky Papers: Jansky Papers, Wisconsin Historical Society, Madison, Wis.

Mem/FCC, NA: Interoffice info. memos, FCC records, RG 173, National Archives, Washington, D.C.

Min/NTIA, NA: Minutes of Committees, 1922–49, records of National Telecommunications and Information Administration (NTIA), RG 417, National Archives, Washington, D.C.

NTIA, NA: Records of NTIA, RG 417, National Archives

RD/FCC, NA: General records, Radio Division, records of the Federal Communications Commission (FCC), RG 173, National Archives, Washington, D.C.

RTPB/FCC, NA: Records of RTPB—Planning Board Meetings, 1942–48, Office of Chief Engineer, Office of Engineering and Technology, FCC records, RG 173, National Archives, Washington, D.C.

Sec/FCC, NA: Records of the Office of Secretary, FCC records, RG 173, National Archives, Washington, D.C.

Sec/GRDC, NA: General correspondence of the secretary, general records, Dept. of Commerce, RG 40, National Archives, Washington, D.C.

TV Hist./FCC, NA: Records relating to history of development of television, 1938–65, FCC records, RG 173, National Archives, Washington, D.C.

Preface and Acknowledgments

1. On the need for more studies of the role of the state in history, see Peter B. Evans, Dietrich Rueschemeyer, Theda Skocpol, eds., *Bringing the State Back In* (New York: Cambridge Univ. Press, 1985) and William E. Leuchtenberg, "The Pertinence of Political History: Reflections on the Significance of the State in America," *Journal of American History* 73 (Dec. 1986). On the importance of standards in industry, see Susanne K. Schmidt and Raymund Werle, *Coordinating Technology: Studies in the International Standardization of Telecommunications* (Cambridge: MIT Press, 1998). Also see the discussion of the different historiographic issues examined in this preface in the "Note on Secondary Sources," following this notes section.

2. For an excellent example of a study on the early history of broadcasting that takes into account the role of different individuals and groups in the construction of radio, see Susan J. Douglas, *Inventing American Broadcasting, 1899–1922* (Baltimore: Johns Hopkins Univ. Press, 1987).

3. "Statement by O. M. Caldwell" (press release), Oct. 15, 1928, b. 36, Dellinger Rec. O. M. Caldwell, "What the Broadcasting Re-allocation Is to Accomplish," Sept. 5, 1928, ibid. Unlike some specialists, I avoid making a sharp distinction between the terms *allocation* and *assignment.*

4. James Lawrence Fly, "Regulation of Radio Broadcasting in the Public Interest," *The Annals of the American Academy of Political and Social Science* 213 (Jan. 1941): 103. I avoid using the term *democratic* here in opposition to the concept of technocracy. This is a notable difference between my work and other studies of contemporary regulatory agencies. See esp. Sheila Jasanoff, *The Fifth Branch: Science Advisers as Policymakers* (Cambridge: Harvard Univ. Press, 1990).

5. For a theoretical treatment of this issue in the case of broadcasting, see esp. Thomas Streeter, *Selling the Air: A Critique of the Policy of Commercial Broadcasting in the United States* (Chicago: Univ. of Chicago Press, 1996).

6. See esp. the essays in Merritt Roe Smith and Leo Marx, eds., *Does Technology Drive History? The Dilemma of Technological Determinism* (Cambridge: MIT Press, 1994).

1. Engineering Public Policy for Radio

1. "Report on Radio Broadcasting Stations for the Year 1922," file 1675, b. 137, RD/FCC, NA. Other scholars have reported similar, although slightly different statistics. See Philip T. Rosen, *The Modern Stentors: Radio Broadcasting and the Federal Government, 1920–1934* (Westport, Conn., Greenwood Press, 1980), 7; Douglas, *Inventing Broadcasting* (see preface, n. 2), xv, 299–303; Hugh G. J. Aitken, "Allocating the Spectrum: The Origins of Radio Regulation," *Technology and Culture* 35 (Oct. 1994): 695.

2. For first quote, see Douglas, *Inventing Broadcasting*, 306. For Hoover quote, see Herbert Hoover, "The Broadcasts of Tomorrow," *Popular Science Monthly* (July 1922); reproduced in Glenn A. Johnson, "Secretary of Commerce Herbert C. Hoover: The First Regulator of American Broadcasting, 1921–1928," Ph.D. diss., Univ. of Iowa, 1970, 294–95.

3. Typescript of radio address given by Herbert Hoover, Mar. 26, 1924, 2, b. 489, fldr. "Commerce Papers—Radio; Corres., press releases, 1924 Jan.-Mar.," Hoover Lib.

4. "Statement by Secretary Hoover at Hearing Before the Committee on the Merchant Marine and Fisheries on H.R. 7357, . . . Mar. 11, 1924," file 67032/7, b. 130, Sec/GRDC, NA.

5. Douglas, *Inventing Broadcasting*, 1–239; Hugh G. J. Aitken, *Syntony and Spark: The Origins of Radio* (New York: Wiley, 1976).

6. Douglas, *Inventing Broadcasting*, 240–68.

7. Ibid., 268–88.

8. Hugh G. J. Aitken, *The Continuous Wave: Technology and American Radio, 1900–1932* (Princeton: Princeton Univ. Press, 1985), 302–479, quote on 498; Douglas, *Inventing Broadcasting*, 288–91.

9. Douglas, *Inventing Broadcasting*, 216–20; Aitken, "Allocating the Spectrum," 686–716.

10. Ibid., 216–20.

11. Ibid., 187–220.

12. Ibid., 221–33.

13. Ibid., 233–39. The 1912 Radio Act is reproduced in Erik Barnouw, *A History of Broadcasting in the United States*, vol. 1, *A Tower in Babel* (New York: Oxford Univ. Press, 1966–70), 291–99. On number of radio bills, see Aitken, "Allocating the Spectrum," 705.

14. Douglas, *Inventing Broadcasting,* 233–39, quote on 237; Rosen, *Modern Stentors,* 20–21.

15. Douglas, *Inventing Broadcasting,* 217. Goldsmith and Pupin quoted in "Interference Between Stations: Its Causes, Effects, and Cures," *Wireless Age* (July 1917): 707–12; see also John Stone Stone, "Remarks on Wave Length Regulations," b. 442, coll. 55, ser. 116, Clark Coll.

16. Aitken, "Allocating the Spectrum," 699.

17. Rosen, *Modern Stentors,* 25–27. For general background on the early history of the Department of Commerce, including the Bureau of Navigation and Bureau of Standards, see Marvin R. Bensman, "The Regulation of Radio Broadcasting by the Department of Commerce, 1921–1927," Ph.D. diss., Univ. of Wisconsin, 1969, 46–71.

18. On "corporate liberalism," see esp. the historiographic discussion in Streeter, *Selling the Air* (preface, n. 5), 28–58.

19. "Report of Radio Conference Held at the Department of Commerce," Mar. 30, 1920, 6, 8, 12–15, 20–21, 30, file 1231NR, b. 109, RD/FCC, NA.

20. On the Institute of Engineers, see A. Michal McMahon, *The Making of a Profession: A Century of Electrical Engineering in America* (New York: IEEE, 1984), 127–73. For origins of the Wave-Length Allocation Committee, see John Stone Stone, "Remarks on Wave Length Regulations," May 12, 1916, 4. For Hogan quote, see John Hogan to J. W. Alexander, Apr. 1, 1920, file 1231NR, b. 109, RD/FCC, NA. For Carty statement, see "Report of Radio Conference Held at the Department of Commerce," Mar. 30, 1920, 10.

21. J. W. Alexander to John Hogan, Apr. 6, 1920, file 1231NR, b. 109, RD/FCC, NA. For examples of other officials who did not think the IRE committee would represent all of the industry, see "Report of Radio Conference," 11, 17, 23.

22. For first quote, see "Report of Radio Conference," 10. On role of standards, see Wilbert F. Snyder and Charles L. Bragaw, *Achievement in Radio: Seventy Years of Radio Science, Technology, Standards, and Measurements at the National Bureau of Standards* (Washington, D.C.: U.S. GPO, 1986), 69–97. On the new committee, see John Hogan to J. W. Alexander, Apr. 9, 1920, file 1231NR, b. 109, RD/FCC, NA. On informal contact with IRE, see Alfred Goldsmith to J. W. Alexander, Apr. 20, 1920, file 1231NR, b. 109, RD/FCC, NA. For last quote, see John Hogan to J. W. Alexander, Apr. 1, 1920.

23. Rosen, *Modern Stentors,* 30–31. On industry protests and meeting with Commerce, see A. E. Kennelly to Hoover, May 28, 1921, file 67032/3, b. 130, Sec/GRDC, NA. For statistic on dominance by military, see Alfred N. Goldsmith to Hoover, Dec. 19, 1921, file 67032/3, b. 130, Sec/GRDC, NA.

24. For Hoover quote, see Herbert Hoover to Alfred N. Goldsmith, Oct. 6, 1921, file 67032/3, b. 130, Sec/GRDC, NA. On Goldsmith's role, see Hoover to Goldsmith, Oct. 28, 1921, file 67032/3, b. 130, Sec/GRDC, NA. On Stratton's

role, see S. W. Stratton to Hoover, Nov. 30, 1921, file 67032/3, b. 130, Sec/GRDC, NA. Goldsmith is quoted in Rosen, *Modern Stentors,* 31.

25. "Statement by the Secretary of Commerce at the Opening of the Radio Conference of Feb. 27, 1922," file 67032/31, b. 131, Sec/GRDC, NA. On allocation and station categories as well as sources of interference before the first radio conference, see Bensman, "Regulation of Radio Broadcasting," 80, 100, 116–17.

26. John M. Jordan, *Machine-Age Ideology: Social Engineering & American Liberalism, 1911–1939* (Chapel Hill: Univ. of North Carolina Press, 1994), 122. Ellis W. Hawley, "Herbert Hoover, the Commerce Secretariat, and the Vision of an 'Associative State,' 1921–1928," *Journal of American History* 61 (June 1974): 116–40.

27. On importance of technical experts, see C. M. Jansky, "Report on Radio Telephone Conference," n.d., fldr. 6, b. 7, Jansky Papers. For Hoover and RCA quotations, see "Minutes of Open Meeting of Department of Commerce Conference on Radio Telephony," Feb. 27 and 28, 1922, 30, b. 496, fldr. "Commerce—Radio Conf. Nat. First, Minutes," Hoover Lib. For second quote, see "Statement by Secretary, Opening of Radio Conference, Feb. 27, 1922."

28. For first quote, see "Tentative Report of Department of Commerce Conference on Radio Telephony," b. 496, fldr. "Commerce—Radio Conf. Nat. First, Minutes," Hoover Lib. On Jansky's statement, see C. M. Jansky (untitled report prior to Feb. 1922 radio conference), fldr. 5, b. 7, Jansky Papers. For Hoover quote, see "Statement by Secretary, Opening of Radio Conference." For RCA quote, see "Memorandum of the Radio Corporation of America with Reference to the Tentative Report of the Department of Commerce Radio Telephony Conference," b. 496, fldr. "Commerce—Radio Conf. Nat. First, Reports & Resol.," Hoover Lib.

29. C. M. Jansky, "Report on Radio Telephone Conference."

30. "Radio Telephony Committee," Mar. 1, 1922, fldr. 2, b. 7, Jansky Papers. On FTC action, see also Barnouw, *History of Broadcasting,* vol. 1, *Tower in Babel,* 116–17.

31. "Report of Department of Commerce Conference on Radio Telephony," b. 496, fldr. "Commerce—Radio Conf. Nat. First, Minutes," Hoover Lib. On the distinction between direct and indirect advertising, see Susan Smulyan, *Selling Radio: The Commercialization of American Broadcasting, 1920–1934* (Washington, D.C.: Smithsonian Institution Press, 1994), 70.

32. Rosen, *Modern Stentors,* 40–46—Harbord quoted on 46. For a general history of IRAC, see "The Interdepartmental Radio Advisory Committee" (IRAC) fldr. "History of the Committee, 1948," b. 103, NTIA, NA. See E. M. Webster, "The Interdepartmental Radio Advisory Committee: Its History, Mode of Operation, and Relationship to Other Agencies," *Proceedings of the Institute of Radio Engineers (PIRE)* 33 (Aug. 1945): 495–99.

33. Goldsmith, quoted by a U.S. radio inspector, Dec. 18, 1922, file 1484,

b. 127, RD/FCC, NA. U.S. radio inspector to commissioner of navigation, Jan. 9, 1923, file 1484, b. 127, RD/FCC, NA.

34. Rosen, *Modern Stentors,* 53–54; Bensman, "Regulation of Radio Broadcasting," 103–9.

35. Quote from 1912 Act reproduced in Barnouw, *History of Broadcasting,* vol. 1, *Tower in Babel,* 293. Bensman, "Regulation of Radio Broadcasting," 150–55; William Terrell, chief radio inspector, quoted on 152.

36. For statistics on number of stations, see Dept. of Commerce news release, Mar. 5, 1923, fldr. "Commerce Papers—Radio: Conferences National—Second," Hoover Lib. "Recommendations of the National Radio Committee" (news release), Mar. 24, 1923, b. 496, fldr. "Commerce—Radio Conf. Nat. First, Minutes," Hoover Lib. On U.S. Navy and frequency shift, see Rosen, *Modern Stentors,* 56–58. For a general discussion of the recommendations of the second radio conference, see Bensman, "Regulation of Radio Broadcasting," 158–62.

37. For Goldsmith statement, see "Minutes of the Radio Telephony Conference," Mar. 23, 1923, fldr. 3, b. 7, Jansky Papers. "Statement by John V. L. Hogan of Principles to Guide Administration of Present Radio Law," Mar. 20, 1923, fldr. 3, b. 7, ibid.

38. On actions taken by the Dept. of Commerce following second radio conference, see Bensman, "Regulation of Radio Broadcasting," 160–71.

39. On legislative efforts prior to third conference, see Bensman, "Regulation of Radio Broadcasting," 85–92, 120–36.

40. Quote from "Recommendation of Subcommittee No. 7," Oct. 9, 1924, fldr. 5, b. 7, Jansky Papers. On the different motivations for national radio, see Smulyan, *Selling Radio,* 20, 21–36—"distance fiend" quote is from 11. On the community-building tradition in communications, see Richard R. John, *Spreading the News: The American Postal System from Franklin to Morse* (Cambridge: Harvard Univ. Press, 1995).

41. For Hoover quote, see *Recommendations for Regulation of Radio adopted by the Third National Radio Conference . . . Oct. 6–10, 1924,* 3, 32–33; copy in b. 496, Commerce Papers-Radio: Conference-Natl. 3rd. corres., Hoover Lib. On experiments with wired interconnection by AT&T, see Smulyan, *Selling Radio,* 52–56.

42. For statement on avoiding monopoly by Hoover, see *Recommendations, Third National Radio Conference, Oct. 1924,* 6. For a more detailed analysis of Sarnoff's effort to develop superpower, see Smulyan, *Selling Radio,* 42–48.

43. Aitken, *Continuous Wave,* 482–83; Rosen, *Modern Stentors,* 64–65. On possible connection between Sarnoff's campaign for superpower and RCA's negotiations with AT&T, see Smulyan, *Selling Radio,* 47; for discussion of attempt by GE and Westinghouse to use shortwave radio to interconnect stations, see 48–52.

44. Rosen, *Modern Stentors,* 62–69. On advertising, see Smulyan, *Selling Radio.* For Hoover quote, see typescript of radio address given by Hoover, Mar. 26, 1924,

8–9, b. 489, fldr. "Commerce Papers—Radio; corres., press releases, 1924 Jan.-Mar.," Hoover Lib.

45. On desirability of high power, see *Recommendations, Third National Radio Conference, Oct. 1924,* 14, 17. Alfred Goldsmith, "Highlights of Radio Broadcasting: Picking Up Broadcast Music," 22, b. 139, RD/FCC, NA. C. W. Horn to S. W. Edwards, Aug. 8, 1924, b. 23, ibid. For Hoover quote, see "Statement by Secretary Hoover at Hearings before the Committee on the Merchant Marine and Fisheries on H.R. 7357," 3, b. 130, Sec/GRDC, NA.

46. For Sarnoff and third quotations, see proceedings report, "Sub-Committee No. 3 . . . Third National Radio Conference . . . Oct. 6–10, 1924," 11, 13, b. 496, fldr. "Commerce Papers—Radio: Conference, National—Third (Proceedings)," Hoover Lib. Goldsmith quoted in J. S. Harbord to S. B. Davis, Nov. 17, 1924, b. 139, RD/FCC, NA. On historical concepts, see esp. essays in Smith and Marx, *Does Technology Drive History?* (see preface, n. 6).

47. "Sub-Committee No. 3, Oct. 6–10, 1924," 19–20.

48. Dellinger and Goldsmith quotations are in "Sub-Committee No. 3, Oct. 6–10, 1924," 24–25, 65. For second Goldsmith quote, see Goldsmith, "Highlights," 44. For statement of GE executive, see Martin Rue to Herbert Hoover, Oct. 18, 1924, b. 132, Sec/GRDC, NA.

49. For Sarnoff and Jansky quotations as well as statement by newspaper owners, see "Sub-Committee No. 3, Oct. 6–10, 1924," 13, 20, 34. Citizens Radio Committee to Herbert Hoover, Oct. 1924, file 2678, b. 153, RD/FCC, NA. For quotation of Tennessee citizen, see P. R. Van Frank to William A. Oldfield, Nov. 14, 1924, ibid.

50. "Sub-Committee No. 3, Oct. 6–10, 1924," 13–14.

51. Ibid., 30–32, 35, 37.

52. Ibid., 43–44.

53. Herbert Hoover to Wallace White, Dec. 4, 1924, fldr. 6, b. 7, Jansky Papers. For last Hoover quote, see "Secretary Hoover Reviews Radio Situation" (press release), Feb. 8, 1925, b. 490, fldr. "Commerce Papers—Radio; corres., press releases, 1924 Jan.-Apr.," Hoover Lib.

54. Herbert Hoover to Wallace White, Dec. 4, 1924.

55. McDonald quoted in Rosen, *Modern Stentors,* 78. For historical background, see Johnson, "Secretary of Commerce Herbert C. Hoover: The First Regulator of American Broadcasting, 1921–1928," 171–73.

56. For work toward allocation system and for second Hoover quote, see "Hoover Reviews Radio Situation," 3. For historical background, see Bensman, "Regulation of Radio Broadcasting," 223–26, 247–61. First Hoover quote from "Radio Can't Live on Jazz," *Kansas City Times,* Dec. 23, 1924; copy in b. 490, fldr. "Commerce Papers—Radio; corres., press releases, 1924 Jan.-Mar.," Hoover Lib.

57. "Hoover Reviews Radio Situation," 4.

58. *Proceedings of the Fourth National Radio Conference and Recommendations for Regulation of Radio, . . . Nov. 9–11, 1925,* 6–7.

59. Herbert Hoover, "Radio Problems and Conference Recommendations" (radio address), Nov. 12, 1925, 4–5, b. 496, fldr. "Commerce Papers: Radio—Conference—National Fourth," Hoover Lib. On how radio should be treated differently from public utilities, see *Proceedings, Fourth National Radio Conference, 1925,* 34. On the origin of the "public interest" standard, see Bensman, "Regulation of Radio Broadcasting," 389.

60. *Proceedings, Fourth National Radio Conference, 1925,* 3.

61. Hoover quotations, ibid., 8–9; for example of the views of educational stations, see reprint of resolution of H. Umberger in "behalf of the Department of Agriculture, the farmers, and agricultural colleges using radio" on 11. Jansky quote from C. M. Jansky to Earle M. Terry, Oct. 16, 1925, fldr. 7, b. 7, Jansky Papers.

62. Owen Young to Herbert Hoover, Dec. 2, 1925, b. 496, fldr. "Commerce Papers: Radio—Conference—National Fourth," Hoover Lib. See also *Proceedings, Fourth National Radio Conference, 1925,* 23.

63. For statistics on power of stations, see ibid., 3, and Bensman, "Regulation of Radio Broadcasting," 247–51. For quote, see "Material on High-Power Broadcasting for Transmittal to House and Senate Committees on Radio Legislation," Jan. 11, 1926, 4, 29, file 1732NR, b. 139, RD/FCC, NA. On RCA trying to connect its support of farmers to university stations broadcasting to rural areas, see "Radio: A Force for Agricultural Progress, A Digest of Opinion Among Agricultural Colleges (RCA)," b. 396, coll. 55, ser. 96, Clark Coll.

64. Aitken, *Continuous Wave,* 484–86; Smulyan, *Selling Radio,* 59.

65. Rosen, *Modern Stentors,* 80–83.

66. On Hoover's response to being called a czar, see Hoover to Karl Broadley, n. d., b. 490, fldr. "Commerce Papers-Radio; Corres., Releases, 1925 May-Sept.," Hoover Lib. Rosen, *Modern Stentors,* 83–85.

67. Rosen, *Modern Stentors,* 85, 93–95; Bensman, "Regulation of Radio Broadcasting," 307–25.

68. *Springfield Republican* editorial (Apr. 18, 1926) in b. 490, fldr. "Commerce Papers-Radio; Corres., Releases, 1925 May-Sept.," Hoover Lib. For general discussion of congressional activity, see Rosen, *Modern Stentors,* 94–100—Pupin quote on 98.

69. On independent regulatory commissions, see esp. Dorothy Nelkin, "Technology and Public Policy," in *Science, Technology, and Society: A Cross-disciplinary Perspective,* ed. Ina Spiegel-Rösing and Derek J. de Solla Price (Beverly Hills, Calif.: Sage, 1977), 422. Adams is quoted by Streeter, *Selling the Air,* 51. For the historical background to Progressive Era attempts to reform the economy, see Robert Britt Horwitz, *The Irony of Regulatory Reform: The Deregulation of American Telecommunications* (New York: Oxford Univ. Press, 1989), 60–69.

70. For Dill's comments, see *Congressional Record,* 69th Cong., 1st Sess., June 30, 1926, 12377; on Senator King's comment that commissioners should be experts, see ibid., 12375.

71. Rosen, *Modern Stentors,* 100–102. For attorney general's evaluation, see William J. Donovan to Hoover, July 8, 1926, b. 490, fldr. "Commerce Papers-Radio; Corres., Releases, 1925 May-Sept.," Hoover Lib. On the amount of "chaos" in this period, see Bensman, "Regulation of Radio Broadcasting," 330–36.

72. Rosen, *Modern Stentors,* 100–106. For general discussion of various legislative efforts following fourth conference, see also Bensman, "Regulation of Radio Broadcasting," 276–79, 339–64. For discussion on possibility of establishment of property rights in the spectrum, see Aitken, "Allocating the Spectrum," 712–13; quote from 713. The *Oak Leaves* case was *The Tribune Company v. Oak Leaves Broadcasting Station, Inc., Coyne Electrical School, Inc., and J. Louis Guyon,* Circuit Court of Cook County, Illinois.

73. Radio Act of 1927 reproduced in Barnouw, *History of Broadcasting,* vol. 1, *Tower in Babel,* 301–5.

74. Alfred N. Goldsmith, "Cooperation between the Institute of Radio Engineers and Manufacturers' Associations," *PIRE* 16 (1928): 1070.

75. Aitken, "Allocating the Spectrum," 709.

76. Ibid., 706–9, 713.

2. Radio Engineers, the Federal Radio Commission, and the Social Shaping of Broadcast Technology

1. For a more limited discussion of the technical decisions of the FRC, see esp. Rosen, *Modern Stentors* (see chap. 1, n. 1) and Robert W. McChesney, *Telecommunications, Mass Media, and Democracy: The Battle for the Control of U.S. Broadcasting, 1928–1935* (New York: Oxford Univ. Press, 1993). For an excellent synthetic discussion of the origin of the tradition of not treating the radio spectrum as a private commodity to be allocated through markets and pricing, see Aitken, "Allocating the Spectrum" (see chap. 1, n. 1). Aitken mainly examines the period before the formation of the FRC; he is more interested in exploring broad economic themes than the underlying values supporting technical decisions.

2. On regulatory commissions, see esp. Nelkin, "Technology and Public Policy" (see chap. 1, n. 69), 422. On the 1934 Communications Act, see esp. McChesney, *Telecommunications and Democracy,* 3.

3. John H. Dellinger, "Engineering Problems of [FRC]," typescript of paper in b. 7, Dellinger Papers.

4. See also Jasanoff, *Fifth Branch* (see preface, n. 4), 229.

5. For 1932 statistics, see FRC, *Sixth Annual Report . . . to Congress,* 4–5. By fall 1928, as many as three engineers attended some commission meetings; by contrast, only one member of the legal division participated at the meetings. See FRC minutes for Oct. 12, 1928, vol. 1, reel 1.1, b. 1, FRC minutes, Sec/FCC, NA.

On statistic from 1929, see FRC, *Third Annual Report,* 7. For quote, see John H. Dellinger, "Radio," typescript of paper in b. 7, Dellinger Papers.

6. For a biographical sketch of Dellinger, see "Obituary," *PIRE* 85 (1963): 794 and Snyder and Bragaw, *Achievement in Radio* (see chap. 1, n. 22), 781–95. On Dellinger's involvement with the 1919 Paris conference and his efforts to explain the government's view to U.S. engineers, see Herbert Hoover to Alfred N. Goldsmith, Oct. 28, 1921, file 67032/3, b. 130, Sec/GRDC, NA. On his early involvement (in 1923) with the interdepartmental committee, see fldr. "Papers of the Interdepartmental Advisory Committee on Government Radio Broadcasting, 1921–1922," b. 18, Dellinger Rec. On Dellinger as chair of the allocation committee of the fourth conference, see *Proceedings, Fourth National Radio Conference* (see chap. 1, n. 58), ii. For examples of Dellinger's popular articles and speeches, see b. 7, Dellinger Papers. On Dellinger's membership with the International Union of Scientific Radiotelegraphy, see b. 439, ser. 116, Clark Coll.

7. In addition to Dellinger, another employee of the Radio Division, Dr. C. B. Jolliffe, was authorized to work for the commission in Feb. 1928, and S. C. Hooper was detailed by the U.S. Navy Department to "aid the Commission in its short wave work." See minutes for Feb. 20, 1928, vol. 1, reel 1.1, b. 1, FRC minutes, Sec/FCC, NA. See also George K. Burgess to W. D. Terrell, Feb. 20, 1928, file 300A, b. 23, RD/FCC, NA. On Dellinger's work during 1927, see Dellinger's "diary notes" in b. 87, Dellinger Rec. On the results of the committee of the American Engineering Council, see "A Statement on Engineering Principles Prepared for Presentation to [FRC]," Mar. 30, 1927, ibid., b. 36. On Dellinger's role becoming official, see FRC, *Third Annual Report,* 12. For an example of Dellinger's effort to organize the Engineering Division and institutionalize close cooperation with the Radio Section, see Dellinger, "Memorandum to the Chairman [FRC]," Aug. 8, 1928, b. 128, file 20-2, ExDir/FCC, NA. On the transfer of the Radio Division of the Dept. of Commerce to the FRC in 1932, see FRC, *Seventh Annual Report* (1933), 1. On Mar. 20, 1928, Dellinger helped organize a meeting of "technical representatives of the government." See b. 128, file 20-2, ExDir/FCC, NA.

8. "Diary Notes," b. 87, Dellinger Rec. For comment of secretary of FRC, see Carl H. Butman, "Memorandum to Judge Robinson," b. 128, file 20-2, ExDir/FCC, NA.

9. On the July 1927 event, see W. H. G. Bullard to secretary of IRE, July 26, 1927, "Radio Advisory Committee" fldr., b. 32, Dellinger Rec. On advisory role of IRE being insitutionalized, see "Report of Broadcast Allocation Committee of the Institute of Radio Engineers to [FRC]," Apr. 6, 1928, ibid., b. 36. On series of meetings in spring-summer 1928, see FRC, *Second Annual Report* (1928), 2–3, 13.

10. On status and membership of broadcast committee during fall 1929, see John H. Dellinger to Alfred N. Goldsmith, Oct. 22, 1928, "IRE Broadcast Committee" fldr., b. 37, Dellinger Rec. See also L. M. Hull (chairman of the committee) to Goldsmith, Nov. 17, 1928, ibid. For Goldsmith quote and for second

quote, see Goldsmith to Dellinger, Oct. 30, 1928, ibid. On committee soliciting advice, see "Brief of the Institute of Radio Engineers method for assisting [FRC]," Dec. 11, 1928, ibid. For statistic on number of engineers involved in each report, see R. H. Marriott, "Work on the Special Radio Broadcasting Problem for [FRC]," Mar. 26, 1929, ibid.

11. John H. Dellinger to Julius Weinberger, Jan. 21, 1929, "RMA, NEMA, NAB" fldr., b. 37, Dellinger Rec. Alfred N. Goldsmith to Dellinger, Jan. 24, 1929, ibid. It should also be noted that Goldsmith was a member of the RMA technical committee. On the effort by the IRE to distance itself from these trade associations, see also McMahon, *Making of a Profession* (see chap. 1, n. 20), 155–56.

12. On Whittemore's involvement, see FRC minutes for Mar. 30, 1928, vol. 1, reel 1.1, b. 1, FRC minutes, Sec/FCC, NA. On Dellinger's role and the use of John V. L. Hogan, L. E. Whittemore, C. M. Jansky, R. S. McBride, and Edgar Felix as "temporary technical advisors," see FRC, *Second Annual Report,* 2–3, 13. On Whittemore's important role during spring 1928, see FRC minutes for Apr. 11, 1928, vol. 1, reel 1.1, b. 1, FRC minutes, Sec/FCC, NA. See also Carl H. Butman, "Memorandum for Judge Sykes," Mar. 30, 1928, b. 128, file 20-2, ExDir/FCC, NA. On Dellinger and Whittemore working together on the reallocation, see also Dellinger's "diary notes" for spring 1928 in b. 87, Dellinger Rec.

13. On Marriott's role with IRE Broadcast Committee and his consulting work for the commission during 1928, see John H. Dellinger to R. H. Marriott, Nov. 10, 1928, "IRE Broadcast Committee" fldr., b. 37, Dellinger Rec. On the plans leading up to Marriott's employment, see also Dellinger to Alfred N. Goldsmith, Nov. 1, 1928, ibid. On Marriott's work with the IRE and FRC, see Dellinger, "Memorandum to the Chairman," Dec. 8, 1928, ibid. On Marriott's consulting work during 1928 and 1929, see Marriott to Dellinger, Jan. 23, 1929, ibid. On Marriott's work with the Bar Association, see "Tentative Report of the Committee on Radio Law of the American Bar Association," Jan. 15, 1929, ibid., b. 36. And see Marriott, "Work on the Special Radio Broadcasting Problems for [FRC]," "IRE Broadcast Committee" fldr., ibid., b. 37. On the consulting work for the commission of the fourth member of the original IRE broadcast committee, John V. L. Hogan, see Carl H. Butman, "Memorandum for Judge Robinson," Jan. 18, 1928, b. 128, file 20-2, ExDir/FCC, NA. See also Butman, "Memorandum for Judge Sykes," Mar. 30, 1928. One important way that major IRE engineers influenced the regulation of radio was through membership on the Radio Advisory Committee (RAC) of the Radio Section, Bureau of Standards. See, for example, "Minutes of Meeting of [RAC]," Mar. 10, 1925, "[RAC]" fldr., b. 32, Dellinger Rec.

14. On the origins of the "public interest" standard, see John H. Dellinger, "Engineering Aspects of the Work of [FRC]," *PIRE* 17 (1929): 1330. See also McChesney, *Telecommunications and Democracy,* 18–37. For quotations of FRC principles, see FRC, *Second Annual Report,* 166–69.

15. FRC, *Second Annual Report,* 55, 168.

16. Herbert Hoover, "Statement by Secretary, Opening of Radio Conference, Feb. 27, 1922," b. 131, file 67032/31, Sec/GRDC, NA. For FRC comment on advertising, see FRC, *Second Annual Report,* 168–69.

17. On the engineering division strongly recommending that licenses of technically inferior stations "be immediately suspended by the Commission," see Guy Hill, "Memorandum to [FRC]," Nov. 26, 1928, b. 23, file 300A, RD/FCC, NA. On Dellinger recommending that "stations should be held rigidly to engineering requirements," see Dellinger to Gen. Saltzman, Oct. 16, 1929, b. 36, Dellinger Rec. On new technical regulations in 1930, see FRC, *Fifth Annual Report* (1931), 24. For an example of testimony by engineers, see the text of the hearing evaluating the application, in 1931, of KLPM in Minot, N.D., docket 1018, b. 201, Dock/FCC, NA.

18. S. E. Frost, *Education's Own Stations: The History of Broadcast Licenses Issued to Educational Institutions* (Chicago: Univ. of Chicago Press, 1937), 4; S. E. Frost, *Is American Radio Democratic?* (Chicago: Univ. of Chicago Press, 1937), 218; and McChesney, *Telecommunications and Democracy,* 30–31. For view of critics that FRC was biased against educational radio, see esp. the publications of Nat. Committee on Education by Radio—for example, Tracy F. Tyler, ed., *Radio as a Cultural Agency: Proceedings of a National Conference on the Use of Radio as a Cultural Agency in a Democracy* (Washington, D.C.: National Committee on Education by Radio, 1934).

19. On the development of NBC, see Barnouw, *History of Broadcasting,* vol. 1, *Tower in Babel* (see chap. 1, n. 13), 145, 181–88. On the growth of CBS, see Rosen, *Modern Stentors,* 149. For comment on "oligopoly of networks," see Frost, *Is American Radio Democratic?* 81. See also McChesney, *Telecommunications and Democracy,* 29.

20. For first quote, see Barnouw, *History of Broadcasting,* vol. 1, *Tower in Babel,* 238–39. See also McChesney, *Telecommunications and Democracy,* 29–30; Smulyan, *Selling Radio,* 65–92. For last quote, see Frost, *Is American Radio Democratic?* 142.

21. "The March of Radio," *Radio Broadcast* 16 (Mar. 1930): 258.

22. McChesney, *Telecommunications and Democracy,* 260–62. For a useful, although much less detailed, discussion of some of the technical policies of the FRC, see also 252–55.

23. The Dept. of Commerce had also made plans for a reallocation, during 1926, in the event Congress passed the appropriate legislation. See acting commissioner to Alfred N. Goldsmith (with newspaper clipping), Feb. 15, 1926, b. 24, file 307-2-NR, RD/FCC, NA. On FRC and the "best scientific opinion," see FRC, *Annual Report* (1927), 1. On Davis Amendment, see McChesney, *Telecommunications and Democracy,* 20–21; Rosen, *Modern Stentors,* 128–32. The amendment is reprinted in FRC, *Second Annual Report,* 11.

24. John H. Dellinger, "Explanation of Allocation of Aug. 13," Aug. 13, 1928, b. 37, "Allocation" fldr., Dellinger Rec. For second Dellinger quote, see Dellinger,

"Analysis of New Broadcast Station Allocation," Sept. 14, 1928, FRC, *Second Annual Report,* 216. "Report of Radio Engineers to [FRC]," *Proceedings of the Institute of Radio Engineers (PIRE)* 16 (1928): 556. "Reports of I.R.E. Committee on Broadcasting," *PIRE* 18 (1930): 15.

25. On district service stations, see Dellinger, "Explanation of Allocation of Aug. 13." "Resolutions Adopted by Conference of Engineers on Apr. 6, 1928," FRC, *Second Annual Report,* 131.

26. John H. Dellinger, "Discussion of Proposals Made at [FRC] Hearing of Apr. 23, 1928," b. 36, Dellinger Rec.

27. On Dellinger's critical commentaries, see Dellinger, "Memorandum to Broadcasting Committee," Aug. 11, 1928, ibid., b. 37, "Allocation" fldr. On provisions of General Order 40, see FRC, *Second Annual Report,* 17, 49–50. See also John H. Dellinger, "Analysis of Broadcasting Station Allocation," *PIRE* 16 (1928): 1477–85. On final proposal forcing stations to share time, see "Statement by Dr. John H. Dellinger" (press release), Sept. 4, 1928, b. 36, Dellinger Rec. On effects of later adjustments to allocation, see "Basis Established by [FRC] for the Division of Radio Broadcast Facilities within the United States," *PIRE* 18 (1930): 2040.

28. Docket 1028, b. 203, Dock/FCC, NA.

29. For Dellinger quote, see "Summary of Discussion at Conference of Engineers on Apr. 6, 1928, by Dr. John H. Dellinger," FRC, *Second Annual Report,* 131. For a historical overview of this ends-means problem, see Leo Marx, "Does Improved Technology Mean Progress?" in *Technology and the Future,* ed., Albert H. Teich, 5th ed. (New York: St. Martin's Press, 1990), 3–14. For the classic critical view of *"technique,"* see Jacques Ellul, *The Technological Society* (New York: Knopf, 1964).

30. For other works analyzing the emphasis on engineering and efficiency as well as the cultural authority of science and technology, see Note on Secondary Sources. For discussions of the related theme of technological determinism, see the essays in Smith and Marx, *Does Technology Drive History?* (see preface, n. 6).

31. McChesney, *Telecommunications and Democracy,* 131.

32. Louis Caldwell, "Radio and the Law," in *Radio and Its Future,* ed. Martin Codel (London: Harper & Bros., 1930; reprint, New York: Arno Press, 1971), 227, 241. For Caldwell's comments on Davis Amendment, see Louis Caldwell, "The Standard of Public Interest, Convenience, or Necessity as Used in the Radio Act of 1927," *Air Law Review* 1 (1930): 316. On Caldwell warning against science being "shackled by legislation," see "Tentative Report of the Committee on Radio Law of the American Bar Association," 15, b. 36, Dellinger Rec. For comments of assistant general counsel, see Paul M. Segal, "The Radio Engineer and the Law," *PIRE* 18 (June 1930): 1038, 1040–41.

33. For Caldwell on "principles of engineering," see "The Problem of Radio Reallocation," *Congressional Digest* 7 (1928): 271. On Caldwell's response to Mor-

gan, see Levering Tyson, ed., *Radio and Education: Proceedings of the First Assembly of the National Advisory Council on Radio in Education, 1931* (Chicago: Univ. of Chicago Press, 1931), 147.

34. On O. M. Caldwell's importance, see Caldwell to members of FRC, July 26, 1928, b. 128, file 20-2, ExDir/FCC, NA. For Caldwell's comments, see "Statement by O. M. Caldwell" (see preface, n. 3); O. M. Caldwell, "What the Reallocation Is to Accomplish" (see preface, n. 3). FRC press release, Nov. 9, 1928, b. 36, Dellinger Rec. "Radio Engineer Analyzes New Broadcasting-Allocation Plan" (press release), Sept. 7, ibid. For press comments that picked up on this theme of the legitimating role of technical experts on the commission, see "The Problem of Radio Reallocation, *Congressional Digest,* 277.

35. Bethuel Webster, "Memorandum for Judge Sykes," Dec. 6, 1929, b. 129, file 20-2, ExDir/FCC, NA. For specific examples of the engineering division reaching conclusions on its own, see esp. division memo to the commissioners in b. 128 and b. 129, file 20-2, ExDir/FCC, NA. On Webster, see also McChesney, *Telecommunications and Democracy,* 32–33, 83–85.

36. For Dellinger and boundary work, see John H. Dellinger, May 26, 1928, b. 52, Min/NTIA, NA. Dellinger's public statements about this boundary were sometimes inconsistent; see "Discussion of the Proposals by Dr. John H. Dellinger," FRC, *Second Annual Report,* 144. For testimony of expert witness, see docket 1019, 27, b. 201, Dock/FCC, NA.

37. On "closure," see for example introductory discussion in Roger Smith and Brian Wynne, eds., *Expert Evidence: Interpreting Science in the Law* (New York: Routledge, 1989), 13. On WSUI, see docket 1028, b. 203, Dock/FCC, NA.

38. Rosen, *Modern Stentors,* 218. On WCAC, see Frost, *Education's Own Stations,* 71–72. Tracy F. Tyler, *Some Interpretations and Conclusions of the Land-Grant Radio Survey* (Washington, D.C.: National Committee on Education by Radio, 1933), 25. On WHAZ, see Rosen, *Modern Stentors,* 168.

39. Rosen, *Modern Stentors,* 260–61. On WCAJ, see Frost, *Education's Own Stations,* 241.

40. Donald MacKenzie, "Missile Accuracy: A Case Study in the Social Process of Technological Change," in Wiebe E. Bijker and John Law, eds., *Shaping Technology/Building Society: Studies in Sociotechnical Change* (Cambridge: MIT Press, 1992), 198. Langdon Winner, "Techne and Politeia: The Technical Constitution of Society," in *Controlling Technology: Contemporary Issues,* ed. William B. Thompson (Buffalo, N.Y.: Prometheus Books, 1991), 299.

41. For statistics on networks, see Rosen, *Modern Stentors,* 146. Ira E. Robinson, "Memorandum for the Commission," Oct. 16, 1928, b. 129, file 20-2, ExDir/FCC, NA. McChesney, *Telecommunications and Democracy,* 35–37.

42. Thomas Hughes, "The Evolution of Large Technological Systems," in Bijker, Hughes, and Pinch, eds., *Shaping Technology/Building Society,* 51. Langdon Winner, *Autonomous Technology: Technics-Out-of-Control as a Theme in Political Thought*

(Cambridge: MIT Press, 1977), 226–51. On technological style see Thomas P. Hughes, *Networks of Power: Electrification in Western Society, 1880–1930* (Baltimore: Johns Hopkins Univ. Press, 1983) 404–60. On the economic consolidation of the 1920s, see Barnouw, *History of Broadcasting,* vol. 1, *Tower in Babel,* 203. On the effect of the depression on radio broadcasting, see McChesney, *Telecommunications and Democracy,* 261.

43. Edwin T. Layton, *The Revolt of the Engineers: Society Responsibility and the American Engineering Profession* (Baltimore: Johns Hopkins Univ. Press, 1986), 4. For an example of the inevitability argument, see David F. Noble, *America by Design: Science, Technology, and the Rise of Corporate Capitalism* (New York: Knopf, 1977).

44. Lee de Forest, "Inaugural Address of Dr. Lee de Forest," *PIRE* 18 (1930): 1123. Lee de Forest, "Statement Concerning Broadcasting," Apr. 7, 1932, roll 4, Lee de Forest Papers, Library of Congress, Washington, D.C.

45. De Forest, "Inaugural Address," 1123. Stanford C. Hooper, "A Spokesman for the Radio Engineer," *PIRE* 19 (1931): 1846.

46. McChesney, *Telecommunications and Democracy,* 156–57.

47. John H. Dellinger, "Radio Paradise" (typescript of talk broadcast over radio on Oct. 31, 1925), b. 7, Dellinger Papers.

48. On early IRE members leaving the AIEE, see Layton, *Revolt of the Engineers,* 43. For 1920 comments of IRE president, see John V. L. Hogan to J. W. Alexander, Apr. 1, 1920, b. 109, file 1231NR, RD/FCC, NA.

49. On Hogan working for broadcast stations, see [?] to G. C. Furness, Nov. 14, 1929, b. 36, Dellinger Rec. On Hogan being asked to advise the FRC on clear-channel stations, see Dellinger to L. M. Hull, Dec. 6, 1928, "IRE Broadcast Committee" fldr., b. 37, ibid.

50. For examples of Goldsmith's involvement with government regulation before 1927, see "Report of Special Radio Conference, held on Jan. 11, 1924," b. 439, ser. 116, Clark Coll.; Alfred N. Goldsmith to S. B. Davis, May 11, 1925, b. 139, file 1732NR, RD/FCC, NA; Goldsmith to S. B. Davis, Jan. 28, 1926, b. 24, file 307-2-NR, ibid; Alfred N. Goldsmith, "The Allocation of Wavelengths to Prevent Interference," reprint from *Year-Book of Wireless Telegraphy and Telephony, 1925,* b. 23, file 307-A-NR, ibid. On Goldsmith's influence on the FRC, see McChesney, *Telecommunications and Democracy,* 283, n. 45. On the RCA executive requesting that Goldsmith assist the government, see John W. Elwood to Goldsmith, Feb. 1, 1922, b. 435, ser. 116, Clark Coll. For Goldsmith's position on advertising, see Rosen, *Modern Stentors,* 163. On Goldsmith's request to the IRE Broadcast Committee, see Robert H. Marriott to Dellinger, Dec. 1, 1929, "IRE Broadcast Committee" fldr., b. 37, Dellinger Rec. On Goldsmith emphasizing cooperation between the IRE and manufacturers, see Goldsmith, "Cooperation" (see chap. 1, n. 74) : 1065–71. On Goldsmith's 1925 claim, see Goldsmith, *High-*

lights of Radio Broadcasting (1925 pamphlet published by NBC), 44, b. 139, file 1732NR, RD/FCC, NA.

51. On Dellinger's sympathy with commercial interests, see memo, "Communications Conference," 21 Oct. 1922, b. 130, file 67032/2, GenCor/GRDC. On the Dept. of Commerce under Hoover, see Rosen, *Modern Stentors,* 61. On Sarnoff and the American Engineering Council, see "A Statement on Engineering Principles prepared for presentation to [FRC]," Mar. 30, 1927.

52. Alfred N. Goldsmith, quoted in "Interference" (see chap. 1, n. 15): 708. On the retarding effect of political solutions, see John Stone Stone, "Remarks on Wave Length Regulations," b. 442, ser. 116, Clark Coll. Michael Pupin, quoted in "Interference": 712. On the early political and economic commitments of IRE members, see also McMahon, *Making of a Profession,* 152–53.

53. C. W. Horn to S. W. Edwards, Aug. 8, 1924, b. 23, file 307NR, RD/FCC, NA. On the parallel development of superpower in electric-power generation and radio broadcasting, see McMahon, *Making of a Profession,* 158–60. Alfred N. Goldsmith to William Brown, June 3, 1924, b. 44, ser. 121, Clark Coll.

54. "Material on High-Power Broadcasting for Transmittal to House and Senate Committees on Radio Legislation," 4, 29, 30, Jan. 11, 1926, b. 139, file 1732NR, RD/FCC, NA. On RCA and rural broadcasting, see, e.g., J. G. Harbord (president of RCA), "Radio and the Farmer," Sept. 16, 1925, b. 139, file 1732NR, ibid. Alfred N. Goldsmith to S. B. Davis, Feb. 16, 1925, b. 139, file 1732NR, ibid. Robinson, quoted by McChesney, *Telecommunications and Democracy,* 35.

3. Competition for Standards

1. Joseph H. Udelson, *The Great Television Race: A History of the American Television Industry, 1925–1941* (University: Univ. of Alabama Press, 1982), 136–37. On estimate of the number of television sets, see Robert H. Stern, *The Federal Communications Commission and Television: The Regulatory Process in an Environment of Rapid Technical Innovation* (New York: Arno, 1979), 185.

2. For a discussion of the process of FCC rule making, see Robert L. Hilliard, *The Federal Communications Commission: A Primer* (Stoneham, Mass: Focal Press, 1991).

3. Udelson, *Great Television Race,* 11–19.

4. Ibid., 24–40. Stern, *FCC and Television,* 91, 96–100.

5. Ibid., 20–24.

6. Albert Abramson, *Zworykin, Pioneer of Television* (Chicago: Univ. of Illinois Press, 1995), 6–61.

7. Ibid., 62–86.

8. Frank C. Waldrop and Joseph Borkin, *Television: A Struggle for Power* (New York: Morrow, 1938), 177–98; Abramson, *Zworykin,* 75–97; Aitken, *Continuous Wave* (see chap. 1, n. 8), 506–11.

9. Martin Codel, "RCA Television Impresses Radio Industry," *Broadcasting,* 11

(Nov. 15, 1936): 10. "Statement on Television by David Sarnoff, President, RCA Corp.," annual stockholders meeting, N.Y.C., May 7, 1935, 6, b. 1765, vol. 3, docket 5806, Dock/FCC, NA. For second Sarnoff quote, see "Sarnoff Urges Against Radio Shakeup," *Broadcasting* 9 (Dec. 1, 1935): 13.

10. "RCA Progress in Television Brings Orders for Equipment," *Broadcasting* 12 (Apr. 15, 1937): 38; "New RCA 441-line Television Proving Impressive in Tests, Asserts Mr. Lohr," *Broadcasting* 12 (May 1, 1937): 34. "Statement by David Sarnoff, President, [RCA] to [RMA], Oct. 20, 1938," b. 103, Dellinger Rec. Abramson, *Zworykin,* 87–158; Sarnoff quoted on 157.

11. Stern, *FCC and Television,* 158.

12. Udelson, *Great Television Race,* 96–120; Abramson, *Zworykin,* 64–170; George Everson, *The Story of Television: The Life of Philo T. Farnsworth* (New York: Arno, 1974).

13. Abramson, *Zworykin,* 104–5, 123; Udelson, *Great Television Race,* 119–21.

14. Martin Codel, "Philco Discloses Its Television Program," *Broadcasting* 11 (Aug. 15, 1936): 10, 55. Philco suit quoted in Udelson, *Great Television Race,* 121.

15. "NBC Rebuilding Television Layout; Two Firms Offer Video Sets to Public," *Broadcasting* 14 (June 15, 1938): 18. The preferred spelling is *DuMont.* I have used it throughout, including within quotations.

16. "NBC Rebuilding Television Layout," 18. Abramson, *Zworykin,* 158–59. Sarnoff quoted ibid., 157. "Is Opposed by McDonald," *Broadcasting* 15 (Nov. 15, 1938): 20.

17. "Including Latest Laboratory Features," *Broadcasting* 15 (Oct. 1, 1938): 6. Udelson, *Great Television Race,* 121–22. On CBS buying RCA equipment, see Martin Codel, "Future of Television Now Up to the Public," *Broadcasting* 15 (Nov. 1, 1938): 14.

18. Udelson, *Great Television Race,* 122–24.

19. "DuMont Video Interest Acquired by Paramount for Movie Experiments," *Broadcasting* 15 (Aug. 15, 1938): 77; "Television Activity Is Spurred as Paramount Acquires Rights," ibid.: 17. On Paramount and DuMont partnership, see also Timothy Reynolds White, "Hollywood's Attempt to Appropriate Television: The Case of Paramount Pictures (DuMont Laboratories)," Ph.D. diss., Univ. of Wisconsin–Madison: 1990. For speculation on relationship between television and the movie industry, see also "Television to Develop Own Art Form after Borrowing First from Movies," *Broadcasting* 12 (June 15, 1937): 30.

20. Udelson, *Great Television Race,* 40–41.

21. Ibid., 41–42.

22. Stern, *FCC and Television,* 64–88; Udelson, *Great Television Race,* 42–45, quote from 43. The 1934 Communications Act is reproduced in Barnouw, *History of Broadcasting* (see chap. 1, n. 13), vol. 2, *The Golden Web,* 311–47.

23. Udelson, *Great Television Race,* 45–46.

24. On new members of television committee, see Donald G. Fink, ed., *Tele-*

vision Standards and Practice: Selected Papers from the Proceedings of the National Television System Committee and Its Panels (New York: McGraw-Hill, 1943), 4–5. "Television Display Impressive to FCC," *Broadcasting* 10 (Jan. 1, 1936): 26.

25. Martin Codel and Sol Taishoff, "Opening of New Radio Frontiers Portrayed," *Broadcasting* 11 (July 1, 1936): 38. "Progress in Ultra-High Band Is Keynote of FCC Hearing," *Broadcasting* 10 (June 1, 1936): 24.

26. Fly quoted in "Hearing to Effect Broadcast Future," *Broadcasting* 10 (May 15, 1936): 45. Craven quoted in "NAB Seeks Long Waves," *Broadcasting* 11 (July 1, 1936): 42.

27. "Notice of Informal Engineering Hearing before the Commission en banc on June 15, 1936," in FCC proceedings report, "Informal Engineering Conference . . . Allocation of Frequencies above 30,000 kc and the Review of Present Frequency Allocations," June 15, 1936, vol. 1, b. 922, docket 3929, Dock/FCC, NA. "Craven Report Outlines Technical Radio Needs," *Broadcasting* 10 (Apr. 15, 1936): 39. On meeting with Roosevelt, see, "Progress in Ultra-High Band Is Keynote of FCC Hearing," 24. FCC, *Second Annual Report . . . Fiscal Year Ended June 30, 1936* (Washington, D.C., 1936), 33–34.

28. For first and fourth quotations, see testimony of William S. Paley and Samuel E. Darby (representing independent radio-set manufacturers) in "Informal Engineering Conference . . . Allocation of Frequencies above 30,000 kc," June 16, 1936, vol. 1, b. 922, 161, 217, docket 3929, Dock/FCC, NA. For second quote, see NAB managing director James W. Baldwin quoted in "Baldwin Asks More Bands," *Broadcasting* 11 (July 1, 1936): 40. For third quote and Baldwin's "American Democracy," see "Samuel E. Darby," *Broadcasting* 11 (July 1, 1936): 40, 116. Robins quoted in "Informal Engineering Conference . . . Allocation of Frequencies above 30,000 kc," June 25, 1936, vol. 3, b. 924, 1832, docket 3929, Dock/FCC, NA.

29. FCC, *Second Annual Report*, 63. Testimony of C. B. Jolliffe for RCA in FCC proceedings report, "Informal Engineering Conference . . . Allocation of Frequencies above 30,000 kc," June 15, 1936, 1259. Testimony of Farnsworth Television, ibid., June 24, 1936, vol. 2, b. 923, 1259, docket 3929, Dock/FCC, NA. On coordination between FCC and RMA engineers, see "Television Service on Everyday Basis by 1938 Forecast," *Broadcasting* 11 (Dec. 1, 1936): 50.

30. "Television Service on Everyday Basis by 1938 Forecast," 50. On RMA's failure to agree on a code of standards as late as June 1938, see "Visual Standards Deferred by RMA," *Broadcasting* 14 (June 15, 1938): 18.

31. Minutes, meeting of Committee on Television (RMA), Sept. 9, 1938, b. 103, Dellinger Rec. E. W. Engstrom to W. R. G. Baker, Sept. 13, 1938, b. 103, ibid.

32. Stern, *FCC and Television*, 144–45. *Milwaukee Journal* quoted by Udelson, *Great Television Race*, 147; see also 126–29.

33. FCC press release, Apr. 18, 1939, fldr. "Television: FCC Rulings, 1937–39," b. 9, TV Hist./FCC, NA.

34. "First Report of Television Committee (FCC): Standards," May 22, 1939, 2–3, 8–9, 13, vol. 2, b. 1763, docket 5806, Dock/FCC, NA.

35. "First Report: Standards," May 22, 1939, 11–15.

36. Howard P. Segal, *Technological Utopianism in American Culture* (Chicago: Univ. of Chicago Press, 1985). David Sarnoff, "Probable Influences of Television on Society," *Journal of Applied Physics* 10 (July 1939): 431. Dunlap quoted by Udelson, *Great Television Race,* 140.

37. "Second Report of Television Committee (FCC)," Nov. 15, 1939, 1–4, vol. 2, b. 1763, docket 5806, Dock/FCC, NA. Stern, *FCC and Television,* 158.

38. Ibid., 8.

39. Ibid., 8–10, 15.

40. FCC press release ("Hearing on Proposed Television Rules"), Dec. 22, 1939, b. 1763, docket 5806, Dock/FCC, NA. For Philco's position at the January meeting, see "Memorandum in Support of Exceptions of Philco Radio and Television Corporation before the FCC . . . Rules and Regulations for Television Stations Tentatively Adopted by the Commission Dec. 21, 1939, and the Specific Recommendations of Its Television Committee," Jan. 29, 1940, vol. 3, b. 1764, docket 5806, ibid. For DuMont's position, see "Brief for Allen B. DuMont Laboratories, Inc.," idem.

41. FCC report, "Order No. 65 Setting Television Rules and Regulations for Further Hearing," May 28, 1940, vol. 7, b. 1768, docket 5806, Dock/FCC, NA.

42. "Brief, DuMont Laboratories," 10.

43. "Memorandum in Support of Exceptions of Philco," Jan. 29, 1940, 20.

44. "Brief, DuMont Laboratories," Jan. 29, 1940, 10–13. "Memorandum in Support of Exceptions of Philco," Jan. 29, 1940, 11–19.

45. FCC report, "Exceptions to the Rules and Regulations for Television Stations Tentatively Adopted by the Commission on Dec. 21, 1939," Feb. 29, 1940, 5, b. 1765, docket 5806, Dock/FCC.

46. Fly and FCC quotations from FCC report, "Exceptions to the Rules . . . Tentatively Adopted . . . Dec. 21, 1939," Feb. 29, 1940, 1 and 4. Wozencraft quoted in FCC report "Order No. 65 Setting Television Rules and Regulations for Further Hearing," May 28, 1940, 14.

47. For an evaluation of Fly's activities on the FCC, see Barnouw, *History of Broadcasting,* vol. 2, *Golden Web,* 173. Fly comments and quotations in FCC proceedings report, "Exceptions to the Rules . . . Tentatively Adopted . . . Dec. 21, 1939," Jan. 19, 1940, 617–21, 637, 667, b. 1762, docket 5806, Dock/FCC, NA.

48. Fly and Baker quoted in FCC, "Exceptions to the Rules . . . Tentatively Adopted . . . Dec. 21, 1939," Jan. 19, 1940, 636, 638, 648.

49. Bond Geddes (RMA executive vice president) to A. S. Wells (RMA president), Jan. 26, 1940, b. 1769, docket 5806, Dock/FCC, NA. For Sarnoff's views, see "Proceedings of the Board of Directors Meeting, [RMA]," Feb. 8, 1940, 4, 14–15, b. 1769, docket 5806, ibid. On commercial television committee, see

meeting minutes, "RMA Committee on Television and RMA Subcommittee on Television Allocations," Jan. 12, 1940, vol. 3, b. 1764, docket 5806, ibid.

50. For Fly's views, see Bond Geddes to A. S. Wells, Jan. 24, 1940, b. 1769, docket 5806, ibid. For Philco statement, see meeting minutes, "RMA Subcommittee on Television Standards," Feb. 29, 1940, b. 1769, docket 5806, ibid.

51. Stern, *FCC and Television,* 168–74.

52. FCC statement of Mar. 22 quoted in Bond Geddes to "RMA Directors and Members," Mar. 25, 1940, b. 1769, docket 5806, Dock/FCC, NA. FCC report, "Order No. 65 Setting Television Rules and Regulations for Further Hearing," May 28, 1940, 18–19.

53. For first two quotations, see article by David Lawrence in *Washington Star* (Mar. 25, 1940) republished in *Congressional Record,* 76th Cong., 3rd Sess., vol. 86, part 14 (app.), Mar. 26, 1940, 1675. *New York Herald Tribune* editorial (Apr. 6, 1940) republished in *Congressional Record,* 76th Cong., 3rd Sess., vol. 86, part 14 (app.), Apr. 8, 1940, 1922. Article by Lawrence in *Washington Star* (Mar. 25, 1940), 1676. On Lawrence as a critic of New Deal and FCC, see "No Legal Power to Scrutinize Programs Is Vested in FCC, Says David Lawrence," *Broadcasting* 9 (Aug. 1, 1935): 40. Letter from Commissioner T. A. M. Craven to Sen. Ernest Lundeen, Apr. 4, 1940, republished in *Congressional Record,* 76th Cong., 3rd Sess., vol. 86, part 14 (app.), Apr. 5, 1940, 1887.

54. *New York Times* editorial (Apr. 8, 1940) republished in *Congressional Record,* 76th Cong., 3rd Sess., vol. 86, part 14 (app.), Apr. 9, 1940, 1968. Letter from Commissioner Craven to Sen. Lundeen, Apr. 4, 1940, republished ibid.

55. *Hearings on S. Res. 251, Requesting Committee on Interstate Commerce to Investigate Actions of Federal Communications Commission in Connection with Development of Television,* Committee on Interstate Commerce, U.S. Senate, 76 Cong., 1st Sess., Apr. 10 and 11, 1940, 2. "Television Escaping Lab Again," *Business Week,* Apr. 20, 1940, 22.

56. For Sarnoff and the first Fly statement, see *Hearings on S. Res. 251,* Apr. 10 and 11, 1940, 30, 53. For second Fly quote, see Fly, *Television* (address over Mutual Broadcasting System and Red Network of the NBC, Apr. 2, 1940), fldr. "Television History, 1939–40," b. 9, TV Hist./FCC, NA. For Philco quote, see supplemental memo "In Behalf of Philco Radio and Television Corporation . . . Sec. 4.73 (b) of the Television Rules and Regulations," May 3, 1940, vol. 7, b. 1765, docket 5806, Dock/FCC, NA.

57. *Hearings on S. Res. 251,* Apr. 10 and 11, 1940, 34–35.

58. "Recommendations of Dr. Lee de Forest and U. A. Sanabria Concerning Television Commercialization," 1–2, Apr. 8, 1940, b. 1769, docket 5806, Dock/FCC, NA.

59. For Roosevelt's support of FCC at press conference, see Stern, *FCC and Television,* 184.

60. FCC report, "Order No. 65 Setting Television Rules and Regulations for Further Hearing," May 28, 1940, 23, 25.

61. On organization of NTSC and different kinds of participants, see "[NTSC] to Speed Development of Uniform Standard" press release, July 17, 1940, fldr. "[NTSC], 1940, 1 of 2," b. 4, TV Hist./FCC, NA. On central coordinating committee, see W. R. G. Baker to Paul J. Larsen, July 24, 1940, ibid. "Statement by E. K. Jett," FCC chief engineer, at NTSC organization meeting, July 31, 1940, ibid. On members of the engineering department organizing demonstrations, see for example K. A. Norton, memo to chief engineer, "Suggestions for [NTSC]," July 29, 1940, ibid. For last quote, see Craven to Lundeen, June 5, 1940, reprinted in *Congressional Record,* 76th Cong., 3rd Sess., app., vol. 86, pt. 16, 3786.

62. Baker quoted in "FCC and RMA Cooperate on National Television" (RMA news bulletin 197), July 31, 1940. "[NTSC]: Scope of Panels," fldr. "[NTSC], 1940, 2 of 2," b. 4, TV Hist./FCC, NA. On statistics of panels, see Donald G. Fink, "What's Happened to Television?: Troubles in Video Engineering and How They Are Busying the Remarkable 'Committee of 169,'" *Technology Review* (Jan. 1941): 1; in records of DuMont Labs., fldr. "NTSC, 1941," b. 88, Library of Congress, Washington, D.C.

63. On first demonstration in August to chairman Fly, see "CBS Announces Television in Color" (CBS news release), Aug. 30, 1940, fldr. "[NTSC], 1940, 1 of 2," b. 4, TV Hist./FCC, NA.

64. Fink, *Television Standards and Practice.* CBS televised live color images in January 1941; see Albert Abramson, *The History of Television, 1880–1941* (Jefferson, N.C.: McFarland 1987), 269.

65. FCC proceedings report, "[NTSC]," Jan. 27, 1941, 2387, b. 1771, docket 5806, Dock/FCC. Fink quoted by Udelson, *Great Television Race,* 155. See also Donald G. Fink, "Perspectives on Television: The Role Played by the Two NTSC's in Preparing Television Service for the American Public," *Proceedings of the IEEE* 64 (Sept. 1976): 1327. For discussion of voting on Mar. 8, see minutes, "Fifth Meeting of [NTSC]," Mar. 8, 1941, fldr. "[NTSC], 1940, 1 of 2," b. 4, TV Hist./FCC, NA.

66. Charles Jolliffe representing RCA quoted in "Television Stalled?" *Business Week,* Mar. 29, 1941, 42. "FCC Television Report, Order, Rules, and Standards," May 3, 1941, 2, b. 1771, docket 5806, Dock/FCC, NA.

67. Udelson, *Great Television Race,* 136–37, 156; Abramson, *History of Television,* 272. On RCA's change in attitude and possible reasons, see also "Television Stalled?" 42. On number of stations broadcasting during the war, see Kittross, "Television Frequency Allocation," 110–11.

68. "Remarks made by J. S. Knowlson at RMA Meeting in New York, 7/31/40," fldr. "[NTSC], 1940, 1 of 2," b. 4, TV Hist./FCC, NA.

69. On report of panel two and Goldsmith quote, see FCC proceedings report, "[NTSC]," Jan. 27, 1941, 2280, 2282. On list of kinds of technical experts

on panel two, see Alfred N. Goldsmith to E. K. Jett, Aug. 1, 1940, fldr. "Apr. 11, 1940–Aug. 1941," b. 2, Mem/FCC, NA. On origins of panel two, see "Second Conference of the Chairmen of the Panels of [NTSC]," Sept. 16, 1940, 1, fldr. "[NTSC], 1940, 1 of 2," b. 4, TV Hist./FCC, NA. On policy of IRE, see minutes, "Institute of Radio Engineers Technical Committee on Television," Apr. 1, 1940, b. 103, Dellinger Rec.

70. For May report, see "FCC Television Report, Order, Rules and Standards," May 3, 1941. For comments during March hearings, including Baker quote, see FCC proceedings report, "[NTSC]," Mar. 20, 1941, 2518, 2544, b. 1771, docket 5806, Dock/FCC, NA.

71. Fred D. Williams, assistant to the president of Philco, to J. S. Knowlson, July 26, 1940, fldr. "[NTSC], 1940, 1 of 2," b. 4, TV Hist./FCC, NA. W. R. G. Baker to F. D. Williams, Aug. 7, 1940, fldr. "[NTSC], 1940, 1 of 2," b. 4, ibid.

72. For first quote and last Philco quote about "two equally good ways," see minutes, "Second Meeting of [NTSC]," Sept. 17, 1940, 4, fldr. "[NTSC], 1940, 1 of 2," b. 4, ibid. For second Philco quote, see Fred D. Williams to W. R. G. Baker, Sept. 5, 1940, ibid. For last quote, see "Memorandum for [DuMont Labs.] on Commercialization," FCC television hearing of Mar. 20, 1941, b. 1771, docket 5806, Dock/FCC, NA.

73. For first quote, see "Conference of the Chairmen of the Panels of [NTSC]," Aug. 21, 1940, 2, fldr. "[NTSC], 1940, 1 of 2," b. 4, TV Hist./FCC, NA. For comments about difficulty in evaluating patent interests, see minutes, "Second Meeting of [NTSC]," Sept. 17, 1940, 3–5.

74. For comments about FCC instructions, see minutes, "Second Meeting of [NTSC]," Sept. 17, 1940, 3–5. For Baker and Fly exchange, see FCC proceedings report, "[NTSC]," Mar. 20, 1941, 2518.

75. For a general discussion of Fly's troubles during this period, see Barnouw, *History of Broadcasting*, vol. 2, *Golden Web*, 173. For examples of the extent of the controversy over this report during the first few months of 1941, see "Fly Foresees Early Decision on Network Monopoly Report," *Broadcasting* 20 (Jan. 20, 1941): 26; "Networks Seek Way to Halt FCC Action," *Broadcasting* 20 (May 12, 1941): 17.

76. For evaluation of panel one and for different responses to request about evaluations of color system, see "Proceedings of Panel No. 1, [NTSC]," vol. 8, b. 1767, docket 5806, Dock/FCC, NA. For comments of panel seven, see FCC proceedings report, "[NTSC]," Jan. 27, 1941, 2306, 2399, 2411. For last quote, see NTSC panel two report, "Proceedings of Panel No. 2 of the NTSC," 22, vol. 8, b. 1767, docket 5806, Dock/FCC, NA.

77. For Baker and Goldsmith quotations, see FCC proceedings report, "[NTSC]," Jan. 27, 1941, 2293, 2302–3. For second quote, see ibid., Mar. 20, 1941, 2558, 2708.

78. On committee requiring majority vote, see "[NTSC]," July 10, 1940, fldr.

"[NTSC], 1940, 1 of 2," b. 4, TV Hist./FCC, NA. For DuMont quote, see "Memorandum for [DuMont Labs.] on Commercialization," 6.

79. Robert Britt Horwitz, *The Irony of Regulatory Reform: The Deregulation of American Telecommunications* (New York: Oxford Univ. Press, 1989), 168, 171.

4. "Rainbow in the Sky"

1. Statement of W. R. G. Baker in "Broadcasters Pledge Action on Post-War Allocation Plans," *Broadcasting* 25 (Aug. 9, 1943): 10. C. M. Jansky, "FM—Educational Radio's Second Chance—Will Educators Grasp It?" n.d., 8–9, fldr. "Jansky and Bailey," b. 124, Armstrong Papers. Armstrong prediction in Edwin H. Armstrong, "Evolution of Frequency Modulation," *Electrical Engineering* 59 (1940): 4. FM commercial operations were suspended during World War II; Armstrong repeated his prediction toward the end of the war. See Armstrong, "The Postwar Future of Broadcasting," address before NAB Executive War Conference, Aug. 31, 1944, in fldr. "EHA—NAB Speech, 1944," b. 18, Armstrong Papers.

2. For historical comparison of AM and FM listening audiences, see Andrew F. Inglis, *Behind the Tube: A History of Broadcasting Technology and Business* (Boston: Focal Press, 1990), 144. "Individual warrior" quote is from [?] Moore to C. B. Fisher, May 27, 1954, fldr. "Misc. papers from Mrs. Armstrong's Files," b. 477, Armstrong Papers. Don V. Erickson, *Armstrong's Fight for FM Broadcasting: One Man vs. Big Business and Bureaucracy* (University: Univ. of Alabama Press, 1973). Lawrence Lessing, *Man of High Fidelity: Edwin Howard Armstrong* (Philadelphia: Lippincott, 1956), 260. A brief account that closely follows Lessing's analysis and fails to use archival sources is Lawrence D. Longley, "The FM Shift in 1945," *Journal of Broadcasting* 12 (1968): 353–64.

3. Armstrong thought the FM shift delayed the "progress of FM broadcasting by more than two years." See Armstrong to secretary, President's Communications Policy Board, Feb. 26, 1951, fldr. "USG—President's Communications Policy Board," b. 470, Armstrong Papers. "Hidden forces" quote is from Armstrong to Charles W. Tobey Jr., Aug. 28, 1950, b. 452, ibid. Comment about engineering mistakes is in Armstrong to E. K. Jett, Dec. 5, 1950, fldr. "E. K. Jett," b. 124, ibid. Armstrong to Charles W. Tobey, Jan. 2, 1946, fldr. "Docket 6651 . . . re Zenith," b. 458, ibid.

4. On Armstrong's early development of FM, see esp. Lessing, *Man of High Fidelity*, 193–223.

5. Ibid., 146. On Armstrong's version of the attempt to block FM, especially the use of a "talk down" campaign, see, e.g., "Statement by Edwin H. Armstrong on Some Ancient History of Radio Art," n.d., fldr. "EHA—NAB Atlantic City," b. 18, Armstrong Papers.

6. On Armstrong's first public demonstration, see Edwin H. Armstrong, "A Method of Reducing Disturbances in Radio Signaling by a System of Frequency Modulation," *PIRE* 24 (1936): 689. FCC, *First Annual Report . . . to Congress . . . Fis-*

cal Year 1935, 28. Memo "Concerning the Activities of Dr. C. B. Jolliffe," May 8, 1948, fldr. "Jolliffe, Dr. C. B.," b. 124, Armstrong Papers.

7. "Statement Concerning Broadcasting Presented by Dr. C. B. Jolliffe [on behalf of RCA and NBC]," FCC hearing on frequency allocation, June 15, 1936, b. 924, vol. 3, docket 3929, Dock/FCC, NA. See also testimony of Charles Jolliffe in *Radio Frequency Modulation: Hearings before the Committee on Interstate and Foreign Commerce on H. J. Res. 78 (A Joint Resolution Relating to Assignment of a Section of the 50-Megacycle Band of Radio Frequencies for Frequency Modulation)*, U.S. HR, 189 Cong., 2d sess. (Feb. 3 and 4, 1948), 248 (hereafter, *Radio Frequency Modulation*).

8. For first quote, see testimony of Edwin H. Armstrong in *Radio Frequency Modulation*, 7. On the May 1936 allocation, see "Text of New FCC Rules Covering Extra-Broadcast Band Services," *Broadcasting* 10 (June 1, 1936): 49. For second quote, see Armstrong to secretary, President's Communications Policy Board, Feb. 26, 1951. During 1936, Armstrong also had difficulty obtaining a license from the FCC to operate his first high-power FM station in Alpine, N.J.; see Erickson, *Armstrong's Fight*, 63–64.

9. For a complete contemporary statement of FM's technical superiority, see "Comparative Potentialities of 'FM' and 'AM' Broadcasting," n.d. (probably 1944), fldr. "Jansky and Bailey," b. 124, Armstrong Papers. See also Inglis, *Behind the Tube*, 119.

10. On FM being boxed in, see Edwin H. Armstrong (untitled memo to the FCC), n.d. (probably 1945), 10, fldr. "EAH Quotes and Misc.," b. 18, Armstrong Papers. In Dec. 1939, one of the commissioners wrote Armstrong that he suspected members of the FCC staff of taking "a supercritical attitude" toward FM. See George Henry Payne to Armstrong, Dec. 8, 1939, fldr. "FCC—Correspondence, 1943," b. 456, ibid. An internal FCC memo gave a favorable evaluation of FM, see "Frequency Modulation vs. Amplitude Modulation—Report on Demonstrations in Schenectady Area," Apr. 29, 1939, fldr. "Apr. 1, 1939–Nov. 30, 1939," b. 1, Mem/FCC, NA. For statistic of 150 stations, see testimony of Armstrong in *Radio Frequency Modulation*, 10. On the early history of FM broadcasters and manufacturers, see Erickson, *Armstrong's Fight*, 64–68.

11. Fly, "Regulation" (see preface, note 4): 103. On disclosure of RCA engineering reports, see Lessing, *Man of High Fidelity*, 242. On Fly's support of FM radio, see also articles and speeches in fldr. "Articles—FM," b. 35, Fly Papers. See also internal FCC memo comparing FM and AM, "General Information [agenda for FCC hearing scheduled to begin Mar. 18, 1940] Aural Broadcasting on Frequencies above 25,000 kc," 4–11, fldr. "Dec. 30, 1939–Mar. 30, 1940," b. 1, Mem/FCC, NA. For statistics of the number of FM stations when the United States entered the war, see FCC, *Seventh Annual Report*, 30. This report also stated (page 30) that in Nov. 1941 there were 150,000 FM receivers in use, with 1,500 being produced every day. In 1943, an FCC commissioner reported, based on "reliable estimates," the 500,000 figure; see E. K. Jett, "Let's Plan Now for Post-

War, Says Jett" *Broadcasting* 25 (Apr. 26, 1943): 30. This was also the prewar number generally accepted by Armstrong and the FM industry. See Armstrong testimony in *Radio Frequency Modulation,* 11. A more recent source states that there were nearly 400,000 FM sets in the hands of the public by the end of 1941. See Inglis, *Behind the Tube,* 129. For the last quote, see Armstrong, untitled address at the fifth annual meeting of FM Broadcasters, Inc., Jan. 26, 1944, 9, fldr. "EHA—FM Broadcasters Address, 1944," b. 18, Armstrong Papers.

12. On wartime stimulation of electronics, see W. R. G. Baker, "Planning Tomorrow's Electronic Highways," *General Electric Review,* 47 (1944): 3. On the establishment of the RTPB, see Baker, "Statement of Operations of the Radio Technical Planning Board," Sept. 28, 1944, fldr. "Radio Technical Planning Board," b. 422, Armstrong Papers. On Fly's important role, see his 1942 speech in fldr. "IRE Rochester Fall Meeting 11–9–42," b. 36, Fly Papers.

13. On the purpose of panel two, see esp. FCC proceedings report, "Allocation of Frequencies to the Various Classes of Non-Governmental Services in the Radio Spectrum from 10 Kilocycles to 30,000,000 Kilocycles," Sept. 28, 1944, vol. 1, 32–33, b. 36, docket 6651, Dock/FCC, NA. Quotations are from Baker, "Statement of Operations of [RTPB]," 3, 7.

14. On the State Department planning for postwar radio, see "Planners Omit International Shortwave," *Broadcasting* 27 (Aug. 14, 1944): 66. A joint meeting of the FCC, State Dept., and IRAC to prepare for future cooperation was held in Nov. 1943; see "Fly Urges Speed in Allocation Studies for FM and Television," *Broadcasting* 25 (Nov. 22, 1943): 12. On IRAC, see also "The Interdepartment Radio Advisory Committee: Its History, Mode of Organization, and Relationship to Other Agencies," Jan. 26, 1945, b. 76, Min/NTIA, NA.

15. For Jansky's involvement with FM, see "Regional FM Allocation Plan Urged," *Broadcasting* 26 (Feb. 14, 1944): 34. FCC 1940 report quoted by Inglis, *Behind the Tube,* 128. Armstrong quote is from "Remarks on Frequency Modulation by Dr. E. H. Armstrong, at Meeting of Technical Subcommittee of [IRAC], Apr. 12, 1937," 6, b. 27, Dellinger Rec.

16. For a discussion of the various types of interference, see "Report of Allocations from 25,000 Kilocycles to 30,000,000 Kilocycles . . . Allocation of Frequencies to the Various Classes of Non-Governmental Services in the Radio Spectrum from 10 Kilocycles to 30,000,000 Kilocycles," May 25, 1945, 49–72, vol. 31, b. 59, ExDir/FCC, NA.

17. "Proceedings, Second Meeting of Panel 5 of the Radio Technical Planning Board," Apr. 11, 1944, 1–57—quote on 25—fldr. "Panel on FM Broadcasting (1 of 2)", b. 1, RTPB/FCC, NA.

18. J. Howard Dellinger to C. M. Jansky, May 1, 1944, fldr. "FCC Hearing—Re: FM Broadcasting," b. 40, Dellinger Rec. For the voting of the engineers, see "[RTPB]: Panel 5, Report on Standards and Frequency Allocations for Postwar

FM Broadcasting," June 1, 1944, 155–64, vol. 22, b. 51, docket 6651, Dock/FCC, NA.

19. "Allocation Conference Opens Friday," *Broadcasting* 27 (Aug. 7, 1944): 16. Description of the FCC Hearings is from Paul A. Porter to Burton K. Wheeler, Mar. 12, 1945, fldr. "Docket 6651—Papers and Correspondence," b. 458, Armstrong Papers. Last quote is from "Allocation of '44," *Broadcasting* 27 (Sept. 25, 1944): 40.

20. On State Dept. conference, see "Allocation Conference Opens Friday," Aug. 7, 1944, 16, and "Allocations Hearings Ordered by FCC," *Broadcasting* 27 (Aug. 21, 1944): 9. Also see minutes of State Dept. meetings in fldr. "World Telecommunications Conference Preparatory Papers," b. 76, Dellinger Rec. For IRAC proposal, see "Proposal of the IRAC for the Revision of Article 7 of the General Radio Regulations (Cairo Revision)," June 15, 1944, fldr. "General Frequency Lists and Proposals—1944 Allocation Hearings," b. 18, Fly Papers. On the flexibility of the State Department, see "Planners Omit International Shortwave," Aug. 14, 1944, 66. See also James P. Veatch, "Memorandum to the Chief Engineer," Aug. 18, 1944, fldr. "July 1, 1944 through Sept. 30, 1944," b. 6, Mem/FCC, NA.

21. FCC proceedings report, "Allocation of Frequencies, Non-Governmental Services, 10 kc to 30,000,000 kc," Sept. 28, 1944, vol. 1, 53–55, 162–65. On how some of the television interests, including DuMont Television and the American Television Society, protested the disruption of their 50–108 band, see page 54 of this report and "RTPB Panel Asks FM, Video Panel," *Broadcasting* 27 (Sept. 11, 1944): 14. Actually, panel two revised its FM recommendation slightly on the last day of the FCC hearing to take into account the needs of amateurs. The final recommendation placed FM in the 43–58 MHz band instead of 41–56 MHz; see "FCC Tackles Conflicting Space Demand," ibid. (Nov. 6, 1944): 11.

22. On the support for the move at the FCC hearings, see "Rapid Growth in High Band Indicated," *Broadcasting* 27 (Oct. 16, 1944): 1. William B. Lodge, "Keeping FM Free from Interference," ibid. (Aug. 14, 1944): 13.

23. Quote is from "Damm Replies to Lodge Article," ibid. (Aug. 28, 1944): 54, 130. On the importance of this kind of competition in the television industry, see also William F. Boddy, "Launching Television: RCA, the FCC and the Battle for Frequency Allocations, 1940–1947," *Historical Journal of Film, Radio and Television* 9 (1989): 45–57.

24. For discussion of companies who wanted to protect the old television system, see "CBS Asks More FM Space, 300 mc Video," *Broadcasting* 27 (Oct. 9, 1944): 9; "New FCC Allocations Seen in Fortnight," ibid. (Oct. 30, 1944): 59; "NAB Panel on FM, Television, and Facsimile," *FM and Television* (1944): 27–28. For Craven testimony, see "Rapid Growth in High Band Indicated": 1.

25. See "Rapid Growth in High Band Indicated": 1; "New FCC Allocations

Seen in Fortnight": 11; "Fly Urges Video in High Frequencies," *Broadcasting* 27 (Sept. 25, 1944): 13.

26. On FCC questioning, see "Rapid Growth in High Band Indicated": 9–10. "Armstrong Defends Wide Band," *Broadcasting* 27 (Oct. 16, 1944): 14.

27. "Interference Data Is Revealed for 30–40 mc Band at Hearing," ibid. (Oct. 23, 1944): 60.

28. On the announcement of the proposal, see "Allocation Proposals Announced by FCC," ibid. 28 (Jan. 16, 1945): 13, 66–67. In Sept. 1944, FCC engineering staff had recommended 150 channels for FM when it proposed shifting FM to the 86–116 MHz band. See "Recommendations on Allocations," Sept. 26, 1944, fldr. "FM—1944 Allocation Hearings—Staff Committee Reports," b. 18, Fly Papers. For the official report on the Jan. 1945 proposal, see "Report of Proposed Allocations from 25,000 Kilocycles to 30,000,000 Kilocycles . . . Allocation of Frequencies to the Various Classes of Non-Governmental Services in the Radio Spectrum from 10 Kilocycles to 30,000,000 Kilocycles," Jan. 15, 1945, b. 31, docket 6651, Dock/FCC, NA.

29. "Allocation Proposals Announced by FCC," Jan. 16, 1945, 67. On reaction of television industry, see also "Reaction Varies to Allocation Proposal," *Broadcasting* 28 (Jan. 22, 1945): 16, 63; and "Reaction to FCC Spectrum Plan," ibid. (Jan. 22, 1945): 59. For first quote, see minutes, IRAC subcommittee on postwar planning in use of radio spectrum, Mar. 9, 1944, 20, fldr. "Meeting of Mar. 9, 1944 (Subcommittee on Postwar Planning)," b. 73, Min/NTIA, NA. For second quote, see minutes, idem, Feb. 24, 1944, 21, fldr. "Meeting of Feb. 24, 1944," ibid. For last quote, see minutes, idem, Mar. 30, 1944, 28, fldr. "Meeting of Mar. 30, 1944," ibid.

30. For first two quotations, see FCC report, "Proposed Allocation, 25,000 kc to 30,000,000 kc," Jan. 15, 1945, 18, 75–76. On the FCC taking into account economic factors, see also memo from FCC Committee 2 to FCC Steering Committee, "Transmittal of Report on Economic Considerations Concerning the FM Industry," Sept. 28, 1944, fldr. "FM—1944 Allocation Hearings—Staff Committee Reports," b. 18, Fly Papers. For an example of technocratic public legitimation of the FCC decision, see testimony of Commissioner Jett before subcommittee of the Committee of Appropriations, HR, on Jan. 18, 1945 in "Extract from Testimony of FCC Commissioner E. K. Jett," fldr. "E. K. Jett," b. 124, Armstrong Papers. For comments of FCC secretary, see T. J. Slowie to Louis Medwin, Jan. 25, 1945, b. 32, docket 6651, Dock/FCC, NA.

31. For an evaluation of Fly's activities on the FCC, see Barnouw, *History of Broadcasting* (see chap. 1, n. 13), vol. 2, *The Golden Web*, 173. "'Flyocracy,'" editorial in *Broadcasting* 26 (Feb. 14, 1944): 40.

32. On congressional investigations of the FCC, see e.g., *Select Committee to Investigate the Federal Communications Commission. Study and Investigation of the Federal Communications Commission: Hearings on H.R. 21*, HR, 78th Cong., 2nd sess.

(1943). Quotations relating charges against the FCC are from "FCC Newspaper Decision before Holidays," *Broadcasting* 25 (Dec. 6, 1943): 7. For a general discussion of Fly's troubles during this period, see Barnouw, *History of Broadcasting,* vol. 2, *Golden Web,* 168–81. Hogan quoted in "What Radio Wants in a Nutshell—Hogan," *Broadcasting* 25 (Dec. 13, 1943): 49. On Dewey, see "Dewey Demands Free Radio, Revised Law," ibid. 27 (Sept. 11, 1944): 11. For a general discussion of government involvement in educational radio, see *FM for Education* (U.S. Office of Education pamphlet), n.d. (1944?), vol. 22, b. 51, docket 6651, Dock/FCC, NA. For the first authorization of educational stations, see FCC, *Fourth Annual Report,* iv, 63. The last quote is from "Last Frontier," editorial in *Broadcasting* 28 (Jan. 22, 1945): 38.

33. Baker quoted in minutes of special IRAC meeting, Dec. 30, 1943, 13, fldr. "Special IRAC Meeting of 30 Dec. 1943," Minutes of Committees, 1922–49, b. 15, Min/NTIA, NA. Second quote from "[RTPB]: Panel 5, Report for Postwar FM Broadcasting," June 1, 1944, 24. Cyril Jansky quoted in FCC proceedings report, "Allocation of Frequencies, Non-Governmental Services, 10 kc to 30,000,000 kc," Sept. 28, 1944, vol. 1, 53.

34. "[RTPB]: Panel 5, Report for Postwar FM Broadcasting," June 1, 1944, 26, 63, 76.

35. For a description of events before and during the February and March hearings, see Paul A. Porter to Burton K. Wheeler, Mar. 12, 1945, fldr. "Docket 6651—Papers and Correspondence," b. 458, Armstrong Papers. On the "secret hearing," see "Synopsis of Proceedings Involving Preparation and Presentation of Radio Industry's Recommendations . . . ," fldr. "[RTPB]," b. 422, ibid.

36. "FM Decision Delayed as FCC Allocates," *Broadcasting* 28 (May 21, 1945): 17. "Allocations Are Unlikely for Fortnight; FCC Said to Favor Wider FM Band," ibid. (May 14, 1945): 17. FCC news release dated May 17, 1945, in fldr. "FCC Hearing—Re: FM Broadcasting," b. 40, Dellinger Rec. A joint committee of industry and FCC engineers was organized to coordinate the observations and analyses. See "FCC Visions FM as Major Radio Service," *Broadcasting* 28 (May 28, 1945): 17. See also "25 Engineers Asked to Assist in FM Tests," ibid. (May 21, 1945): 78; FCC news release, May 18, 1945, in fldr. "FCC—Correspondence, 1945," b. 456, Armstrong Papers.

37. On War Production Board announcement, see "WPB to Lift Construction Ban on V-J Day," *Broadcasting* 28 (June 11, 1945): 15, 72. For views of three trade associations, see "TBA, FMBI Demand Quick Allocation," ibid. (June 4, 1945): 16; "New Hearing on FM Proposals Called," ibid. (June 18, 1945): 16. For quote of Zenith president, see Eugene F. McDonald to Howard Vincent O'Brien, June 12, 1945, fldr. "Docket 6651—Papers and Correspondence," b. 458, Armstrong Papers. On decision following June 22/23 hearing, see "FCC Allocates 88–106 mc Band to FM," *Broadcasting* 29 (July 2, 1945): 13.

38. Memo, "Why Television Needs the Lower Frequency Channels as Pro-

posed by the FCC," Feb. 20, 1945, fldr. "Color Television: Working Papers," b. 3, Col. TV Hist./FCC, NA.

39. FCC proceedings report, "Allocation of Frequencies, Non-Governmental Services, 10 kc to 30,000,000 kc," Jan. 15, 1945, 11, 45, 46. For Norton statement, see "Statement of K. A. Norton," June 22, 1945, fldr. "FCC Hearing—Re: FM broadcasting," b. 40, Dellinger Rec.

40. For first quote, see Paul A. Porter to Burton K. Wheeler, Mar. 12, 1945, file 66-4a, b. 281, ExDir/FCC, NA. See also Porter to Edward A. Kelly, May 4, 1945, b. 281, ibid. Second quote is from Porter to Clyde M. Reed, Mar. 28, 1945, file 66-4a, b. 281, ibid. Testimony of George Sterling in *Radio Frequency Modulation,* 187.

41. For first quote, see Paul A. Porter to E. F. McDonald, Feb. 4, 1946, fldr. "Docket 6651 . . . re Zenith," b. 458, Armstrong Papers. On FCC evaluation, see "FCC Surveying Effects on Receiver Costs of Moving FM Band Upward," *Broadcasting* 28 (Apr. 2, 1945): 79. See also McDonald to Burton K. Wheeler, Mar. 26, 1945, fldr. "Docket 6651—Papers and Correspondence," b. 458, Armstrong Papers. For CBS prediction, see "CBS Predicts FM Will Supplant AM; Promotes Color Video in '45 Report," *Broadcasting* 30 (Apr. 1, 1946): 29. For testimony on conversion, see "Brief of Stromberg-Carlson Company," vol. 2, b. 33, docket 6651, Dock/FCC, NA. Others testified that conversion would take up to five years; see esp. Philco testimony in "Military to Confide Secret Data to Radio," *Broadcasting* 28 (Mar. 13, 1945): 68, 71. On FCC evaluation of this testimony, see untitled FCC report (probably June 27, 1945 report), fldr. "FCC Hearing—Re: FM Broadcasting," b. 40, Dellinger Rec. On the FCC emphasizing availability of converters, see "Shifting of FM Upward in Spectrum Seen," *Broadcasting* 28 (Mar. 19, 1945): 18.

42. On Philco's position, see "Analysis of the Implications, for Educational Broadcasting of [FCC] Report Issued Jan. 16, 1945," Feb. 21, 1945, 15, vol. 2, b. 33, docket 6651, Dock/FCC, NA. For ABC's position, see Frank Marx (director of general engineering), "Statement to [FCC] on the Proposed FM and Television Allocations," Feb. 22, 1945, ibid. For CBS's position, see "Brief of [CBS]," Feb. 20, 1945, ibid. For Cowles Broadcasting, see "Brief Filed in Behalf of [Cowles Broadcasting]," ibid. For Majestic Radio and Television, see "Statement for [FCC] on 'Receiver Design Considerations for Proposed Frequency Modulation Band' by [Majestic]," Feb. 27, 1945, ibid. For position of Hallicrafters, see "FM Allocation to Feature FCC Hearing," *Broadcasting* 28 (Feb. 26, 1945): 13. DuMont Television actually wanted to give most of the VHF band to television and move FM even further upwards. See "DuMont Would Give 44–216 mc to Television, Eliminating FM," ibid. (Mar. 5, 1945): 13. For Crosley's position, see Eugene F. McDonald to Paul A. Porter, Apr. 4, 1945, fldr. "Docket 6651—Papers and Correspondence," b. 458, Armstrong Papers. For position of amateurs and police

chiefs, see "RTPB, FMBI Propose Counter-Allocation," *Broadcasting* 28 (Feb. 5, 1945): 66; "FCC Allocates 88–106 mc Band to FM": 13.

43. On the involvement of the U.S. Office of Education, see "FCC Has Open Mind on FM and Television," *Broadcasting* 28 (Feb. 12, 1945): 15. On the Michigan Commission, see Joseph E. Maddy to T. S. Slowie, Feb. 15, 1945, b. 32, docket 6651, Dock/FCC, NA. For the Journal Company, see "Brief of the Journal Company, Licensee of WMFM, Milwaukee, Wisconsin," vol. 2, b. 33, ibid. On Stromberg-Carlson, see "Brief of Stromberg-Carlson Company." On panel five, see "Brief on Behalf of Panel 5 'FM Broadcasting' of the Radio Technical Planning Board," n.d. (Feb. 1945?), vol. 2, b. 33, docket 6651, Dock/FCC, NA. On the position of GE and Zenith, see Eugene F. McDonald to Paul A. Porter, Apr. 4, 1945. Panel seven (panel on facsimile) of the RTPB also actively opposed the shift; see John V. L. Hogan, "Statement on Behalf of Panel 7," vol. 29, b. 58, ExDir/FCC, NA. For the position of RCA and NBC, see "Brief for Radio Corporation of America, National Broadcasting Company, Inc., RCA Communications, Inc., Radiomarine Corporation of America," vol. 2, b. 33, docket 6651, Dock/FCC, NA. Although FM proponents thought RCA executives privately favored the shift as a way to hurt FM, there is no evidence to support this position. See McDonald to Craven, Apr. 16, 1945, fldr. "Docket 6651—Papers and Correspondence," b. 458, Armstrong Papers. The other FM manufacturers that opposed the shift were Ansley Radio Corp., Espey Manufacturing Co., Freed Radio Corp., Garod Radio Corp., Meissner Manufacturing Co., Pilot Radio Corp., Radio Engineering Labs., and Scott Radio Labs. See "Resolution Prepared at Conference of Pioneer FM Radio Manufacturers, Wednesday, June 6, 1945—Hotel Waldorf Astoria, New York City," b. 34, docket 6651, Dock/FCC, NA.

44. On the effort to construct a unified scientific front, see "RTPB, FMBI Propose Counter-Allocation": 15. On McDonald's contention, see Eugene F. McDonald to Paul A. Porter, Apr. 4, 1945. The eleven engineers who McDonald claimed supported shifting FM were K. A. Norton, Frank Marx of ABC, W. B. Lodge of CBS, John D. Reid of Crosley, Cyrus T. Read of Hallicrafters, David B. Smith of Philco, D. E. Noble of Galvin Manufacturing, T. A. M. Craven of Cowles Broadcasting, T. T. Goldsmith of DuMont Labs., E. W. Allen of the FCC, and Archer S. Taylor of Paul Godley Co. Evidence of other engineers who supported Norton, besides the eleven listed above, can be found in the following sources: for Lewis M. Clement of Crosley Corp., see Clement to R. H. Manson, Feb. 10, 1945, b. 422, fldr. "[RTPB]," b. 422, Armstrong Papers; for radar engineer Edward P. Tilton, see "Shifting of FM Upwards in Spectrum Seen," 18; for H. W. Wells of the Carnegie Institution Dept. of Terrestrial Magnetism, see Porter to McDonald, Mar. 28, 1945, fldr. "Docket 6651—Papers and Correspondence," b. 458, Armstrong Papers; for chief engineer Morris Pierce of WGAR in Cleveland, see McDonald to Craven, Apr. 19, 1945, fldr. "Docket 6651—Papers and Correspondence," b. 458, ibid. For first quote, see Craven to McDonald, Apr. 30, 1945, ibid. On the poll of engineers,

see also "Military to Confide Secret Data to Radio," Mar. 13, 1945, 70. For second quote, see Craven to McDonald, Apr. 10, 1945, fldr. "Docket 6651—Papers and Correspondence," b. 458, Armstrong Papers.

45. For first quote, see "Statement on Radio Propagation: Evidence before [FCC] in docket 6651 by C. M. Jansky, Jr., Chairman of Panel 5, FM Broadcasting, of [RTPB]," Feb. 28, 1945, fldr. "FCC Hearing—Re: FM Broadcasting," b. 40, Dellinger Rec. For second quote, see "Brief on Behalf of Panel 5."

46. For Armstrong quote, see Armstrong, "Memorandum for Senator Charles W. Tobey," Mar. 26, 1948, b. 452, Armstrong Papers. For argument that technical evidence was sufficient, see also Eugene McDonald to Congress, Apr. 20, 1943, fldr. "Docket 6651—Papers and Correspondence," b. 458, ibid.

47. "Exhibit No. 577, Filed on Behalf of Panel 5, FM Broadcasting, of [RTPB]: Memorandum Concerning the Steps which Must Be Taken . . . in any Attempt to Predict Possible Interference with [VHF] Services from F2 Layer Transmission," fldr. "FCC Hearing—Re: FM Broadcasting," b. 40, Dellinger Rec.

48. For first Armstrong quote, see "Supplemental Brief of Edwin H. Armstrong in Opposition to Proposed FM Assignments," Apr. 18, 1945, 16, fldr. "FCC Hearing—Re: FM Broadcasting," b. 40, Dellinger Rec. Armstrong quote on theory and practice is in fldr. "EAH Quotes and Misc.," b. 18, Armstrong Papers. On Armstrong denouncing theoreticians, see Edwin H. Armstrong, "Mathematical Theory vs. Physical Concept," *FM and Television* (1944): 11–13, 36. See also Lessing, *Man of High Fidelity,* 199–200. On Armstrong preferring Beverage's testimony, see "Brief of Edwin H. Armstrong in Opposition to Proposed FM Assignments," Feb. 21, 1945, 8, vol. 2, b. 33, docket 6651, Dock/FCC, NA.

49. Armstrong discussed E-layer and tropospheric effects in his "Brief in Opposition," 15–18. On tropospheric effects, see also "Confidential Brief of Armstrong before FCC," Apr. 25, 1945, fldr. "FCC Hearing—Re: FM Broadcasting," b. 40, Dellinger Rec. On the effort to point out Norton's assumptions, see "Brief on Behalf of Panel 5," 9–23. A good example of nitpicking is a letter to the FCC where Armstrong pursues a lengthy discussion trying to clarify if a question mark should have appeared after a statement in an official transcript. See Armstrong to E. K. Jett, Apr. 23, 1945, fldr. "FCC Correspondence, 1945," b. 456, Armstrong Papers.

50. For first quote, see "Brief on Behalf of Panel 5," 9. For statement of WMFM, see "Brief of the Journal Company." The statement of the chairman of panel seven is from John V. L. Hogan, "Statement on Behalf of Panel 7." The last quote is from "Brief in Opposition," 1945.

51. Edward W. Allen to J. H. Dellinger, Jan. 30, 1945, fldr. "FCC Hearing—Re: FM Broadcasting," b. 40, Dellinger Rec.

52. On 2–1 rejection ratio, see "FCC Allocates 88–106 mc Band to FM," July 2, 1945, 13. On the assumption of a 10–1 rejection ratio, see the text of the FCC

decision to move FM, reproduced in "FCC Allocations Order Text," *Broadcasting* 29 (July 2, 1945): 64.

53. On belief in impropriety of debating FM publicly, see W. J. Damm to W. R. G. Baker, Mar. 6, 1944, fldr. "[RTPB]," b. 422, Armstrong Papers. See also Eugene F. McDonald to FCC commissioners, Nov. 26, 1945, fldr. "Docket 6651— . . . re Zenith," b. 458, Armstrong Papers. On the failure of dual-band FM to gain authorization, see "FCC Denies Zenith Plea for FM Change," *Broadcasting* 30 (Jan. 28, 1946): 15, 91. Paul A. Porter to McDonald, Feb. 4, 1946. FCC report, "Allocation of Frequencies, Non-Governmental Services, 10 kc to 30,000,000 kc," Mar. 5, 1946, vol. 29, b. 58, ExDir/FCC, NA. On "kid gloves" quote, see "Fight Over FM Rural Coverage to Go on Despite FCC Edict," *Heinl Radio News Service,* Jan. 30, 1946, in fldr. "E. K. Jett," b. 124, Armstrong Papers. Amstrong did attempt before the war to publicize FM radio, but the 1945 decision was a major turning point. For an example of Armstrong's efforts, see Edwin H. Armstrong, "The New Radio Freedom," *Journal of the Franklin Institute* 232 (Sept. 1941): 210–18.

54. For first quote, see Millard C. Faught to Edwin H. Armstrong, Feb. 17, 1949, fldr. "Faught Co.," b. 113, Armstrong Papers. On Armstrong hiring PR consultant as early as March 1946, see Armstrong to John Orr Young, Mar. 7, 1946, fldr. "Correspondence between Armstrong and Public Relations Firms," b. 157, ibid. On Armstrong giving ideas to consultants and on consultants sending him articles, see Young to Armstrong, Mar. 25, 1946, ibid. On use of releases, see Young to Armstrong, Mar. 7, 1946, ibid. Millard C. Faught to Armstrong, Apr. 26, 1949, fldr. "Faught Co.," b. 113, ibid. For supportive comments from the editor of *FM and Television,* see Milton B. Sleeper to Armstrong, Oct. 3, 1945, fldr. "Sleeper, Milton B.," b. 427, ibid. On support from *Radio and Television Retailing,* see editor to Armstrong, Oct. 6 [19??], fldr. "An Important Message to the Radio Trade," b. 18, ibid. On $1,000 payments, see Young and Faught to Armstrong, Feb. 28, 1947, fldr. "Correspondence between Armstrong and Public Relations Firms," b. 157, ibid. For examples of speeches, see b. 18, ibid. For examples of Armstrong sending material to newspapers, magazines, engineers, politicians, and broadcasters, see fldr. "EHA—FCC Docket 8487—Heinl News & Other Coverage, Correspondence Regarding Briefs," b. 19, and fldr. "EHA letters to editors," b. 4, ibid. For last quote, see Marion Claire to Armstrong, Sept. 24, 1947, fldr. "EHA-FMA address—Chicago 1947," b. 18, ibid.

55. For first two quotations, see John Orr Young to Eugene F. McDonald, Apr. 4, 1946, fldr. "Faught Co.," b. 113, Armstrong Papers. For comments about "FM's virility," see Ben Strouse (chairman of NAB FM committee) to managers (of FM member-stations), Sept. 22, 1950, fldr. "NAB," b. 136, ibid. See also FMBI booklet publicizing FM in b. 115, ibid. Edwin H. Armstrong, "An Important Message to the Radio Trade Regarding Frequency Modulation Receivers," fldr. "An Important Message . . . " b. 18, ibid.

56. On McDonald convincing others to support letter-writing campaign, see

McDonald to Charles W. Tobey, Apr. 9, 1945, fldr. "Docket 6651—Papers & Correspondence," b. 458, Armstrong Papers. On McDonald involving GE and Stromberg-Carlson, see McDonald to Burton K. Wheeler, Apr. 23, 1945, fldr. "Docket 6651—Papers & Correspondence," b. 458, ibid. On Tobey's role, see Frank J. Sulloway to Armstrong, Oct. 28, 1950, fldr. "Tobey, Charles W.," b. 432; Charles W. Tobey to Armstrong, Sept. 29, 1949, ibid. For quote, see McDonald to Armstrong, Jan. 3, 1946, fldr. "Docket 6651— . . . re Zenith," b. 458, ibid.

57. Eugene F. McDonald to FCC commissioners, Nov. 26, 1945. For first quote, see Edwin C. Johnson, *Congressional Record,* vol. 95, pt. 4, 81st Cong., 1st Sess., Apr. 20, 1949, 4781. For example of congressman writing to FCC to support low-band FM, see Harold H. Hagen to FCC, Apr. 24, 1945, b. 281, ExDir/FCC, NA. For material detailing Armstrong's lobbying efforts during 1947, see fldr. "EHA—FCC Docket 8487," b. 19, Fly Papers. For last quote, see Armstrong's testimony in *Radio Frequency Modulation,* 30.

58. Millard C. Faught, "We Have Lost Our Freedom of Speech," 13, fldr. "Faught Co.," b. 113, Armstrong Papers. On FM and free radio see also Edwin H. Armstrong, untitled radio address, Oct. 15, 1947, fldr. "EHA-FMA address—Chicago," b. 18, ibid. For argument about ability of FM to accommodate hundreds of stations, see Jansky, "FM—Educational Radio's Second Chance—Will Educators Grasp It?" 7–8, fldr. titled "Jansky and Bailey," b. 124, ibid. See also Edwin H. Armstrong, "The New Radio Freedom," *Journal of the Franklin Institute* 232 (Sept. 1941): 214. Eugene F. McDonald to Charles W. Tobey, Apr. 9, 1945. For last point, see [?] to McDonald, Feb. 12, 1946, fldr. "Docket 6651–. . . re Zenith," b. 458, Armstrong Papers.

59. *Radio Frequency Modulation,* 2, 166–77.

60. [?] Moore to C. B. Fisher, May 27, 1954. The last point was emphasized by NBC and RCA. See, e.g., Niles Trammel, "FM: A Statement of NBC's FM Policy," Jan. 21, 1944, b. 104, Dellinger Rec.

61. On the importance of rhetorical strategies and boundary work in science and technology studies, see "Note on Secondary Sources," following this notes section.

Chapter 5. VHF and UHF

1. For statistics, see John Michael Kittross, "Television Frequency Allocation in the United States," Ph.D. diss., Univ. of Illinois, 1959, 295–305. Kittross's dissertation is useful but, unlike this book, does not use extensive archival sources to gain a deeper understanding of the events described.

2. Craven testimony in FCC proceedings report, "Allocation of Frequencies to the Various Classes of Non-governmental Services . . . 10 Kilocycles to 30,000,000 Kilocycles," Oct. 17, 1944, 3494, b. 39, docket 6651, Dock/FCC, NA. Panel eleven claim, ibid., Sept. 28, 1944, 110, b. 36.

3. Kittross, "Television Frequency Allocation," 10. On IRAC denying re-

quests, see E. M. Webster, "The Interdepartment Radio Advisory Committee" (address dated Jan. 26, 1945), 5–7, b. 76, NTIA, NA. On "dangerous precedent," see P. F. Siling to Commissioner Craven, Apr. 28, 1944, fldr. "IRAC Committee, general Correspondence and Reports, 1922–49 (1944)," b. 115, ibid. For last quote, see Robert L. Batts to Daniel E. Noble, Oct. 28, 1944, fldr. "Engineering Department Panel (2) on Frequency Allocation, 1943–44," b. 1, RTPB/FCC, NA.

4. Webster, "[IRAC]" (address), 7. First Craven quote, minutes, special meeting of IRAC, June 15, 1944, 24, fldr. "Report of 1 June 1944 (Special Subcommittee on Postwar Planning)," b. 73, Min/NTIA, NA. Second Craven quote from Senate testimony, quoted in "[IRAC]: The Charges," fldr. "IRAC Outline and Statement," 20, b. 76, ibid.

5. First Craven quote in minutes of meeting of IRAC subcommittee on postwar radio spectrum, Aug. 12, 1943, 3, fldr. "Meeting of Aug. 12, 1943 (Subcommittee on Postwar Planning)," b. 73, Min/NTIA, NA. Minutes of special meeting, IRAC, Dec. 30, 1943, 9, fldr. "Special IRAC Meeting," b. 15, ibid. For second Craven quote, see minutes of special meeting of IRAC, June 15, 1944, 24.

6. Jolliffe claim in minutes, IRAC subcommittee on postwar radio spectrum, Feb. 24, 1944, 21, fldr. "Meeting of Feb. 24," b. 73, Min/NTIA, NA. For statement about reason for needing frequencies in the 250 MHz range, see minutes of second meeting of panel two of RTPB, Sept. 7 and 8, 1944, 12, fldr. "Engineering Department Panel (2) on Frequency Allocation, 1943–44," b. 1, RTPB/FCC, NA. Goldsmith claim in minutes of first meeting of panel one, Dec. 22, 1943, 8, fldr. "Engineering Department Panel (1) on Spectrum Utilization, 1943–44 (2 of 2)," b. 1, ibid. Baker remarks are in minutes of special meeting of IRAC, Dec. 30, 1943, 13.

7. For Armstrong's comments, see "Statement Submitted to the U.S. Senate Commerce Committee," 81 Cong., 1st Sess. on S. 1973, June 17, 1949, 140–42. Craven quote from FCC proceedings report, "Petition of [CBS] for Change in Rules . . . Concerning Television Broadcast Stations," Dec. 13, 1946, 784, b. 802, docket 7896, Dock/FCC, NA. Fly quoted by Kittross, "Television Frequency Allocation," 149. Quotation of W. C. Lent in minutes, IRAC subcommittee on postwar radio spectrum, Mar. 30, 1944, 2, fldr. "Meeting of Mar. 30, 1944," b. 73, Min/NTIA, NA.

8. "Television: A Case of War Neurosis," *Fortune* (Feb. 1946).

9. FCC report, "Proposed Allocation, 25,000 kc to 30,000,000 kc," Jan. 15, 1945, 19, b. 31, docket 6651, Dock/FCC, NA.

10. On order of priority adopted by IRAC, see minutes of special meeting of IRAC, June 15, 1944, 2–3. For a similar list of priorities used by State Dept. committee, see minutes of meeting of Committee 2, State Dept. Special Committee on Communications, Aug. 11, 1944, fldr. "World Telecommunications Conference Preparatory Papers," b. 76, Dellinger Rec. On priorities adopted by FCC, see

FCC report, "Proposed Allocation, 25,000 kc to 30,000,000 kc," Jan. 15, 1945, 19.

11. Craven claim in FCC proceedings report, "Allocation of Frequencies, Non-governmental Services, 10 kc to 30,000,000 kc," Oct. 17, 1944, 3494–95. Testimony of amateurs, ibid., Oct. 6, 1944, 512, 553, b. 37, docket 6651, Dock/FCC, NA.

12. IRAC claim in FCC proceedings report, "Allocation of Frequencies, Non-governmental Services, 10 kc to 30,000,000 kc," Oct. 17, 1944, 3502.

13. On limited knowledge of co-channel interference above 150 MHz and for RTPB quote, see FCC proceedings report, "Allocation of Frequencies, Non-governmental Services, 10 kc to 30,000,000 kc," Oct. 14, 1944, 1695, 1701, b. 38, docket 6651, Dock/FCC, NA. Panel six calculation, ibid., Sept. 28, 1944, 66. For last quote, see "Questions to Be Directed to Mr. Smith," fldr. "Television: 1944 Allocations, 1 of 4," b. 7, TV Hist./FCC, NA.

14. FCC quoted in FCC proceedings report, "Allocation of Frequencies, Non-governmental Services, 10 kc to 30,000,000 kc," Jan. 15, 1945, 8.

15. On historical precedent of conversion of spark transmitters, see testimony of Goldsmith in minutes of second meeting of panel one—Committee 2 of RTPB, Jan. 19, 1944, 1–2, fldr. "Engineering Department Panel (1) on Spectrum Utilization, 1943–44 (2 of 2)," b. 1, RTPB/FCC, NA.

16. Testimony of labor union in FCC proceedings report, "Allocation of Frequencies, Non-governmental Services, 10 kc to 30,000,000 kc," Oct. 17, 1944, 2456–57. "Statement of Norman D. Waters for the American Television Society at the FCC Television Hearing, Washington, D.C.," Oct. 12, 1944. For companies supporting CBS, see FCC report, "Proposed Allocation, 25,000 kc to 30,000,000 kc," Jan. 15, 1945, 90. On CBS's efforts during this period to convince the FCC and the industry to authorize color, see also Bradley Francis Chisholm, "The CBS Color Television Venture: A Study of Failed Innovation in the Broadcasting Industry," Ph.D. diss., Univ. of Wisconsin–Madison, 1987, 175–234.

17. CBS testimony in FCC proceedings report, "Allocation of Frequencies, Non-governmental Services, 10 kc to 30,000,000 kc," Oct. 17, 1944, 3479.

18. Goldmark testimony in minutes, meeting of IRAC subcommittee on postwar radio spectrum, Jan. 20, 1944, 6, fldr. "Meeting of Feb. 10, 1944 (Subcommittee on Postwar Planning)," b. 73, Min/NTIA, NA. On conflicting evaluations of when tubes would be ready, see app. 12 of "Statement for Introduction into the FCC," Oct. 28, 1944, fldr. "Engineering Department Panel (2) on Frequency Allocation, 1943–44," b. 1, RTPB/FCC, NA. "Statement of C. B. Jolliffe . . . at the Frequency Allocations Hearing before the Federal Communications Oct. 1944," fldr. "Television: 1944 Allocations, 1 of 4," b. 7, TV Hist./FCC, NA.

19. C. B. Jolliffe to E. K. Jett, Dec. 22, 1944, fldr. "Engineering Department Panel (2) on Frequency Allocation, 1943–44," b. 1, RTPB/FCC, NA. Jansky com-

ments about AM radio in minutes, meeting of IRAC subcommittee on postwar radio spectrum, Feb. 10, 1944, 3.

20. Baker quote in minutes, special meeting of IRAC, Dec. 30, 1943, 8, 14. Chisholm, "CBS Color Venture," 218.

21. FCC comments in FCC report, "Proposed Allocation, 25,000 kc to 30,000,000 kc," Jan. 15, 1945, 90.

22. FCC quoted in FCC proceedings report, "Allocation of Frequencies, Non-governmental Services . . . 10 kc to 30,000,000 kc," Jan. 15, 1945, 8.

23. First Jett quote ibid., Mar. 1, 1945, 4907, b. 43, docket 6651, Dock/FCC, NA. Second Jett quote ibid., Jan. 15, 1945, 14–15. Testimony of TBA representative, ibid., Mar. 2, 1945, 4958.

24. Philco testimony, ibid., 4899. For comments of TBA, see "Supplementary Exhibit for Television Broadcasters Association," Feb. 1945, fldr. "Television: 1944 Allocations, 4 of 4," b. 7, TV Hist./FCC, NA. For industry views about adequacy of twelve or thirteen channels, see FCC proceedings report, "Allocation of Frequencies, Non-governmental Services, 10 kc to 30,000,000 kc," Mar. 1, 1945, 4899.

25. On instructions to take into account "economic factors," see minutes of meetings in fldr. "Color Television: Working Papers," b. 3, Col. TV Hist./FCC, NA. Jett quoted in FCC proceedings report, "Promulgation of Rules and Regulations and Standards of Good Engineering Practice for Commercial Television Broadcast Stations," Oct. 12, 1945, 203, b. 2313, docket 6780, Dock/FCC, NA.

26. TBA comments in FCC proceedings report, "Promulgation of Rules, Regulations, Standards of Good Engineering Practice," Oct. 11, 1945, 81–84.

27. "Report by the Commission," Nov. 21, 1945, 2, 8, b. 2313, docket 6780, Dock/FCC, NA.

28. Sanabria proposal in FCC proceedings report, "Promulgation of Rules, Regulations, Standards of Good Engineering Practice," Oct. 12, 1945, 287–94. On the possibility of better programming, see the channel-sharing proposal of station WWDC in Washington, D.C., in "Testimony before the [FCC] Television Hearing," Oct. 11, 1945, 2, docket 6780, Dock/FCC, NA.

29. Although the FCC did not mandate channel sharing, it did allow voluntary sharing. See "Report by the Commission," Nov. 21, 1945, 8. The 150- and 75-mile separations were proposed in September; see FCC order 85053, Sept. 1945, b. 2313, docket 6780, Dock/FCC, NA. For first quote, see FCC proceedings report, "Promulgation of Rules, Regulations, Standards of Good Engineering Practice," Oct. 12, 1945, 196–204. For second quote, see "Trial Brief [CBS]," n.d., 5–6, fldr. "Television: 1944 Allocations, 2 of 4," b. 7, TV Hist./FCC, NA.

30. For recommendation of station separations of 170 and 85 miles, see committee minutes in fldr. "Color Television: Working Papers," b. 3, Col. TV Hist./FCC, NA. For first quote, see Robert L. Batts to Daniel E. Noble, Oct. 28, 1944. For second quote, see E. W. Allen, "Very-High-Frequency and Ultra-High-

Frequency Signal Ranges as Limited by Noise and Co-Channel Interference (paper presented at IRE meeting, Jan. 24–27, 1945)," fldr. "UHF and VHF Signal Ranges," b. 40, Dellinger Rec. On conflicting views of tropospheric interference presented during meetings of the RTPB, see various panel six/Committee 4 reports in fldrs. "Allocation Hearing: Committee 4 Television Broadcast" and "Television: 1944 Allocation," boxes 1 and 6, TV Hist./FCC, NA. Goldsmith comments in minutes of meetings in fldr. "Color Television: Working Papers," b. 3, Col. TV Hist./FCC, NA. On RTPB having originally recommended separations with understanding that they did not take into account tropospheric interference, see "Report of Special Committee of Panel 2: Considering Proposals for Allocation Plan in the 60 Megacycle Region," app. 11, 1, Oct. 26, 1944, fldr. "Engineering Department Panel (2) on Frequency Allocation, 1943–44," b. 1, RTPB/FCC, NA. For TBA recommendations of Dec. 1944, see "Television Allocation Plans (handwritten notes)," Dec. 1944, fldr. "Television History, 1928–45," b. 9, TV Hist./FCC, NA. For CBS prediction that shadows and ghosts would not be a problem, see "Statement of Peter C. Goldmark," fldr. "Television: 1944 Allocations, 3 of 4," b. 7, TV Hist./FCC, NA.

31. For first quote, see "Report by the Commission," Nov. 21, 1945, 2. For second and last quotes, see "Standards of Good Engineering Practice Concerning Television Broadcast Stations," Dec. 19, 1945, 8, 11, fldr. "Television History," b. 8, TV Hist./FCC, NA.

32. For statistic on 115 applications, see "Notice to Applicants for Experimental Television Stations and to Applicants for Experimental and Developmental FM Stations," fldr. "Television Allocations: Pertinent Releases, 1946–49," b. 6, TV Hist./FCC, NA. For comments of CBS executive, see FCC proceedings report, "Petition of [CBS] for Change in Rules . . . Concerning Television Broadcast Stations," Dec. 9, 1946, 53, b. 801, docket 7896, Dock/FCC, NA. Chisholm, "CBS Color Venture," 258, 272–77.

33. "Report of [RTPB]. Panel 6 to [FCC]," Nov. 1946, A-3, fldr. 5, b. 26, DuMont Coll. For first quote, see "Statement Presented by W. R. G. Baker," Feb. 10, 1947, 4, fldr. "Docket 7896, Exhibits 58–75," b. 4, Col. TV Hist./FCC, NA. For second quote, see FCC proceedings report, "Petition of [CBS] for Change in Rules . . . Concerning Television Broadcast Stations," Dec. 13, 1946, 681. On studies coordinated by commission, see "Report on Project No. W-26. Field Intensity Measurements at 700 Megacycles" (FCC, Engineering Dept., Technical Information Div.), Dec. 6, 1946, fldr. 5, b. 26, DuMont Coll.

34. For first quote, see "Petition of [CBS] for Color Television Transmission Standards (Docket 7896)," fldr. "Petition Proposing a Revised VHF Plan, Docket 8736, 8975, 9175. 1949," b. 5, TV Hist./FCC, NA. Thomas Goldsmith testimony in FCC proceedings report, "Petition of [CBS] for Change in Rules . . . Concerning Television Broadcast Stations," Feb. 12, 1947, 1754, b. 803, docket 7896,

ibid. Handwritten memo, n.d., fldr. "Petition Proposing a Revised VHF Plan, Dockets 8736, 8975, 9175. 1949," b. 5, TV Hist./FCC, NA.

35. T. A. M. Craven testimony (quotation) in FCC proceedings report, "Petition of [CBS] for Change in Rules . . . Concerning Television Broadcast Stations," Feb. 12, 1947, 1867. Jett testimony ibid., Dec. 12, 1946, 666, b. 801, docket 7896, Dock/FCC, NA.

36. "Statement of Dr. C. B. Jolliffe (Docket 7896)," Dec. 9, 1946, 16, fldr. "Docket 7896, Exhibits 29–31," b. 3, Col. TV Hist./FCC, NA. FCC report, "Petition of [CBS] for Change in Rules . . . Concerning Television Broadcast Stations," Mar. 18, 1947, 4, fldr. "Television Allocations, Pertinent Releases, 1946–49," b. 6, TV Hist./FCC, NA.

37. Westinghouse and Federal testimony in FCC proceedings report, "Petition of [CBS] for Change in Rules . . . Concerning Television Broadcast Stations," Dec. 12, 1946, 593, 642.

38. For June 1947 statistic, see "37 Cities in 24 States Represented by 65 Television Grants and Nine Applicants" (FCC news release), June 5, 1947, fldr. "Television Allocations, Pertinent Releases, 1946–49," b. 6, TV Hist./FCC, NA. See Sept. 1948 data in FCC proceedings report, "Amendment of Section 3.606 of the Commission's Rules and Regulations," Sept. 13, 1948, 1669, b. 63, docket 8736, Dock/FCC, NA. On number of TV sets sold, see Chisholm, "CBS Color Venture," 303.

39. "Notice of Proposed Rule Making . . . Amendments to the Commission's Rules and Regulations Governing Sharing of Television Channels and Assignment of Frequencies to Television and Non-Government Fixed and Mobile Services" (docket 8487), Aug. 19, 1947, fldr. "Television," b. 6, TV Hist./FCC, NA. A committee of the RTPB provided important testimony against sharing. See "Interference to Television Service Resulting from Shared Operations by Fixed and Mobile Services: Report of RTPB Panel 6 Committee on Shared Services," Nov. 5, 1947, fldr. "Sharing of Television Channels . . . Docket 8487, 1947," b. 5, TV Hist./FCC, NA.

40. Testimony of employee of General Telephone Corp. in FCC proceedings report, "Amendments to the Commission's Rules and Regulations Governing Sharing of Television Channels and Assignment of Frequencies to Television and Non-Government Fixed and Mobile Services," Nov. 17, 1947, 77, 84, b. 3357, docket 8487, Dock/FCC, NA.

41. Ibid., Nov. 18, 1947, 336–37, b. 3358.

42. Commissioner Wayne Coy testimony in FCC proceedings report, "Amendment of Section 3.606 of the Commission's Rules and Regulations," Sept. 13, 1948, 1666–68. On agreement with Canada, see "Tentative Canadian-USA Television Allocation Plans," Dec. 10, 1947, fldr. "Television," b. 6, TV Hist./FCC, NA.

43. For example of testimony about tropospheric propagation, see FCC proceedings report, "Amendment of Section 3.606 of the Commission's Rules and

Regulations," July 26, 1948, 1329–42, b. 62, docket 8736, Dock/FCC, NA. Coy testimony ibid., Sept. 13, 1948, 1670, 1673. On need for "adequate planning," see "Report and Order . . . Amendment of Section 3.606 of the Commission's Rules and Regulations" (dockets 8975 and 8736), Sept. 30, 1948, fldr. "Television Allocations, Pertinent Releases, 1946–49," b. 6, TV Hist./FCC, NA.

44. Audrey L. van Dort, "JTAC through the Years," July 1977 (unpublished paper in possession of author), 1.

45. JTAC proceedings, "Utilization of Ultra-High Frequencies for Television: Report of [JTAC] (IRE-RMA) to [FCC]," Sept. 20, 1948, 52–54, b. 267, docket 8976, Dock/FCC, NA.

46. For first quote, see testimony of Philip Siling of the JTAC in FCC proceedings report, "Utilization of Frequencies in the Band 475 to 890 Megacycles for Television Broadcasting," Sept. 20, 1948, 55, b. 266, docket 8976, Dock/FCC, NA. For second quote, see "Utilization of Ultra-High Frequencies for Television: Report of [JTAC] (IRE-RMA) to [FCC]," 55. Testimony of FCC staff members in FCC proceedings report, "Series of Engineering Conferences . . . Amendments of Section 3.606 of the Commission's Rules and Regulations and the Amendment of Rules Concerning Television and Frequency Modulation Broadcasting Services," Dec. 2, 1947, C-291–2, b. 64, docket 8736, Dock/FCC, NA.

47. Testimony of DuMont engineer Thomas Goldsmith in FCC proceedings report, "Utilization of Frequencies in the Band 475 to 890 Megacycles for Television Broadcasting," Sept. 20, 1948, 169–73.

48. JTAC testimony in FCC proceedings report, "Utilization of Frequencies in the Band 475 to 890 Megacycles," 11–12.

49. "Utilization of Ultra-High Frequencies for Television," 1–9, 30, 45–47.

50. JTAC testimony in FCC proceedings report, "Series of Engineering Conferences . . . ," Dec. 2, 1948, C-259. For second quote, see JTAC proceedings, "Allocation Standards for VHF Television and FM Broadcasting," Dec. 2, 1948, 3, b. 68, docket 8736, Dock/FCC, NA. For testimony of Fink and quote about boxing match, see FCC proceedings report, "Series of Engineering Conferences," Dec. 2, 1948, C-284, C-288. On tests with RCA sponsored by the JTAC, see "Engineering Report (Television Co-channel Interference Test for JTAC)," fldr. "Color Television: JTAC & RMA letters re: television demonstrations," b. 3, Col. TV Hist./FCC, NA.

51. For discussion of compromises, see also FCC proceedings report, "Series of Engineering Conferences," Nov. 30, 1948, C-14, b. 67, docket 8736, Dock/FCC, NA.

52. Quote about committee's mandate from FCC, *Fifteenth Annual Report,* 43. For list of members, see attachment A in "Report of the Ad Hoc Committee for the Evaluation of the Radio Propagation Factors Concerning the Television and Frequency Modulation Broadcasting Services in the Frequency Range Between

50 and 250 Mc," vol. 1, May 31, 1949, b. 69, docket 8736, Dock/FCC, NA. FCC, *Fifteenth Annual Report,* 146.

53. Section 4 in "Report of the Ad Hoc Committee," vol. 1, May 31, 1949.

54. Memo from Thomas Carroll to E. W. Allen and members of the ad hoc committee, Jan. 1949, ibid., attachments B-1, B-3.

55. Copy of "[FCC] Notice of Further Proposed Rule Making," July 11, 1949 in fldr. 9, b. 29, ser. 8, DuMont Coll.

56. Edwin C. Johnson to David Sarnoff, May 13, 1949, b. 69, docket 8736, Dock/FCC, NA.

57. Johnson quoted in "Johnson's Stand, Clarifies Position on TV," *Broadcasting* 37 (Nov. 21, 1949): 56. Coy statement quoted in "UHF Band, Power Development is Key—Coy," ibid. 36 (Mar. 28, 1949): 54.

58. Kittross, "Television Frequency Allocation," 226–30. "Additional Views of Commissioner Hennock," included in Wayne Coy to Edwin C. Johnson, Feb. 26, 1949, fldr. "Television Allocations, Pertinent Releases, 1946–49," b. 6, TV Hist./FCC, NA.

59. FCC, *Sixteenth Annual Report,* 102. On DuMont proposal, see "Comments and Exhibits Concerning FCC Dockets 8736, 8976, and 9175, Allen B. DuMont Laboratories," Aug. 26, 1949, fldr. 12, b. 28, DuMont Coll. For views of ABC, see "Statement of Frank G. Kear in Behalf of [ABC]," fldr. 8, b. 30, DuMont Coll.

60. "A National Television Allocation Plan, [DuMont Labs]," Feb. 6, 1950, fldr. 7, b. 29, DuMont Coll.

61. "Report of the Ad Hoc Committee," vol. 2, fldr. 29, b. 29, DuMont Coll. For role of Technical Research Division, see FCC, *Seventeenth Annual Report,* 155–57. On collaboration between Standards and FCC, see also T. J. Slowie to Newbern Smith, Sept. 29, 1950, fldr. "1949 Television Hearing: Misc. Correspondence (2 of 2)," b. 6, TV Hist./FCC, NA. Testimony of Standards staff members in FCC proceedings report, "Amendment of Section 3.606 of the Commission's Rules and Regulations . . .," Oct. 19, 1950, 12953–54, b. 113, docket 8736, Dock/FCC, NA. On collaboration among government agencies and industry, see FCC, *Sixteenth Annual Report,* 150–51. For information about project in Connecticut, see fldr. "RCA-NBC Investigation of UHF Transmission and Reception in the Bridgeport-Stratford, Conn. Area," b. 5, TV Hist./FCC, NA.

62. "Testimony of Dr. Thomas T. Goldsmith, Jr.; [DuMont Labs.]; Oct. 19, 1950," fldr. 9, b. 30, DuMont Coll. For claim by DuMont representative about fulfilling the second priority, see R. P. Wakeman to Thomas T. Goldsmith (DuMont Labs. interoffice correspondence on FCC hearings), Oct. 23, 1950, fldr. 9, b. 30, DuMont Coll. For statistic about number of UHF channels used by two plans, see FCC proceedings report, "Amendment of Section 3.606 of the Commission's Rules and Regulations; Amendment of the Commission's Rules," Oct. 24, 1950, 13600, b. 113, docket 8736, Dock/FCC, NA.

63. On comments of chief of the Technical Research Division (Edward

Allen), see testimony in FCC proceedings report, "Amendment of Section 3.606 of the Commission's Rules and Regulations," Oct. 16, 1950, 12777–78, b. 113, docket 8736, Dock/FCC, NA.

64. On ABC's position, see ibid., Nov. 10, 1950, 13930, b. 113, docket 8736, Dock/FCC, NA. JTAC quote, ibid., Oct. 27, 1950, 14141–42, b. 113, docket 8736, Dock/FCC, NA. For commission's claim about majority of witnesses, see untitled information memo from FCC chief engineer to commissioners, Dec. 13, 1950, 7–8, fldr. "TV Assignment and Planning, 1950–51 (1 of 2)," b. 8, TV Hist./FCC, NA. For last quote, see "TV Topics" (FCC memo from engineering staff to commissioners), n.d., 9, fldr. "TV Assignment and Planning, 1950–51 (1 of 2)," b. 8, TV Hist./FCC, NA.

65. For first quote about stratovision, see FCC proceedings report, "Amendment of Section 3.606 of the Commission's Rules and Regulations," Oct. 24, 1950, 13653–54. Senator Johnson quoted in "Additional Views of Commissioner Jones," 11, included in Wayne Coy to Edwin C. Johnson, Feb. 26, 1949, fldr. "Television Allocations, Pertinent Releases, 1946–49," b. 6, TV Hist./FCC, NA. On commission assuring Johnson, see ibid., 11. For quote of views of the commission's engineers, see "Memorandum," from chief of the FCC Technical Research Division to chief engineer, July 10, 1950, 8, fldr. "Television Assignment and Planning: 1950–51 (2 of 2)," b. 8, TV Hist./FCC, NA; "TV Topics" (FCC memo from engineering staff to commissioners), n.d., 6.

66. FCC proceedings report, "Amendment of Section 3.606 of the Commission's Rules and Regulations," Oct. 24, 1950, 13616.

67. On comments of Office of Education, see ibid., Nov. 27, 1950, 15761–62, 15770, b. 118, docket 8736, Dock/FCC, NA.

68. Ibid., 15770.

69. Hennock quote, ibid. Nov. 29, 1950, 16280, b. 118, docket 8736, Dock/FCC, NA.

70. "Third Notice of Further Proposed Rule Making," fldr. "Television History," b. 8, TV Hist./FCC, NA. "The DuMont 1951 National Television Allocation Plan," May 22, 1951, fldr. 9, b. 30, DuMont Coll.

71. Kittross, "Television Frequency Allocation," 294.

72. "Sixth Report and Order," Apr. 11, 1952, 80–81, b. 118, docket 8736, Dock/FCC, NA.

73. A detailed assignment plan presented to the commission in 1950 by an engineer named Bernard C. O'Brien (a report reasonably well received by the industry) also argued for reducing station separations. See Kittross, "Television Frequency Allocation," 215–17.

74. "Dissenting Opinion of Commissioner Jones," 20–21, 24, 31—included with "Sixth Report and Order," Apr. 11, 1952, b. 118, docket 8736.

75. Ibid., 15.

76. Kittross, "Television Frequency Allocation," 292.

77. Ibid., 314–23.

78. Ibid., 324–32.

6. Competition for Color-Television Standards

1. "CBS Announces Television in Color (CBS news release)," Aug. 30, 1940, fldr. "[NTSC], 1940, 1 of 2," b. 4, TV Hist./FCC, NA. Testimony of Adrian Murphy in FCC proceedings report, "Petition of [CBS] for Changes in Rules . . . Concerning Television Broadcast Stations," Dec. 9, 1946, vol. 1, b. 801, 61, docket 7896, Dock/FCC, NA.

2. For Goldsmith's comments, see Alfred N. Goldsmith to Peter C. Goldmark, Sept. 27, 1940 in "Proceedings of Panel No 2, [NTSC]," b. 1767, docket 5806, Dock/FCC, NA. Testimony of E. Ray Cummings in FCC proceedings report, "Petition of [CBS] for Changes in Rules . . . Concerning Television Broadcast Stations," Dec. 9, 1946, 617.

3. "Television Report, Order, Rules and Standards," May 3, 1941, b. 1771, docket 5806, Dock/FCC, NA.

4. Baker's comments reprinted in annex to "Separate Opinion of Commissioner Jones Dissenting in Part," A-2, in "First Report of Commission (Color Television Issues)," Sept. 1, 1950, fldr. "TV Assignment and Planning: 1950–51 (1 of 2)," b. 8, TV Hist./FCC, NA. On CBS color and the manufacturers, see Bradley Francis Chisholm, "The CBS Color Television Venture: A Study of Failed Innovation in the Broadcasting Industry," Ph.D. diss., Univ. of Wisconsin–Madison, 1987, 142.

5. On role of war in preventing FCC from taking action after six months, see "First Report (Color Television Issues)," Sept. 1, 1950, 6.

6. Chisholm, "CBS Color Venture," 99–100, 186–87.

7. Ibid., 185.

8. Ibid., 201.

9. Ibid., 226–27, 252–63.

10. "First Report (Color Television Issues)," 7.

11. Chisholm, "CBS Color Venture," 263–68. On RCA's 1940 presentation, see "First Report (Color Television Issues)," 4.

12. Chisholm, "CBS Color Venture," 267–72.

13. Jolliffe quoted in "First Report (Color Television Issues)," 8.

14. FCC proceedings report, "Petition of [CBS] for Changes in Rules . . . Concerning Television Broadcast Stations," Feb. 13, 1947, vol. 8, b. 803, 1944, 1949, docket 7896, Dock/FCC, NA. Engstrom statement ibid., Jan. 29, 1947, vol. 8, b. 802, 1133, docket 7896, Dock/FCC, NA. For RCA's views, see "Report of the Commission, March 1947 . . . Petition of [CBS] for Changes in Rules . . . (Docket 7896)," 4–11, fldr. "Color Television: Standards, Preliminary Releases," b. 2, Col. TV Hist./FCC, NA.

15. Handwritten note with Mar. 1947 FCC report, n.d., fldr. "Color Televi-

sion: Standards, Preliminary Releases," b. 2, Col. TV Hist./FCC, NA. "Report of Engineering Department Special Committee No. 1 of [RMA] . . . Questions in Connection with Color Television; Statement Presented by W. R. G. Baker, Chairman, Committee No. 1," Feb. 10, 1947, fldr. 2, ser. 8, b. 27, DuMont Coll.

16. Memo "Re: Petition of [CBS] for Color Television Transmission Standards (Docket 7896)," n.d., 4, fldr. "Petition Proposing a Revised VHF Plan, Dockets 8736, 8975, 9175. 1949," b. 5, TV Hist./FCC, NA. On commissioners demanding a high level of field testing, see "Report of the Commission, March 1947 . . . Petition of [CBS] for Changes in Rules . . . (Docket 7896)," 2–3. On need for healthy safety factor, see "Brightness and Flicker" (short document with Mar. 1947 FCC report), n.d., fldr. "Color Television: Standards, Preliminary Releases," b. 2, Col. TV Hist./FCC, NA. On concern that citizens not be treated as guinea pigs, see memo "Re: Petition of [CBS] for Color Television Transmission Standards (Docket 7896)," n.d., 5.

17. "Popular Support for Standards" (short document with Mar. 1947 FCC report), n.d., fldr. "Color Television: Standards, Preliminary Releases," b. 2, Col. TV Hist./FCC, NA. Handwritten note with Mar. 1947 FCC report, n.d., ibid. Chisholm, "CBS Color Venture," 300–301.

18. "Report of the Commission, March 1947: Petition of [CBS] for Changes in Rules . . . (Docket 7896)," 2–3.

19. Hecht testimony in FCC proceedings report, "Petition of [CBS] for Changes in Rules . . . Concerning Television Broadcast Stations," Feb. 11, 1947, vol. 8, b. 803, 1503, docket 7896, Dock/FCC, NA.

20. Hecht testimony ibid., 1944, 1949, 1529.

21. Hecht testimony ibid., 1514, 1543.

22. RCA testimony ibid., Feb. 13, 1947, 1979. Critical remarks of CBS, ibid., Dec. 12, 1946, b. 801, 561–63, docket 7896.

23. For first quote, see "Genius and the Practical Realities" (short document with Mar. 1947 FCC report), n.d., fldr. "Color Television: Standards, Preliminary Releases," b. 2, Col. TV Hist./FCC, NA. For second quote, see "Brightness and Flicker" (short document, ibid.), n.d. Hecht testimony in FCC proceedings report, "Petition of [CBS] for Changes in Rules," Feb. 11, 1947, 1547.

24. Craven testimony in FCC, "Petition of [CBS] for Changes in Rules," Dec. 13, 1946, b. 802, 791–92, docket 7896, Dock/FCC, NA.

25. Memo "Re: Petition of [CBS] for Color Television Transmission Standards (Docket 7896)," n.d., 1–4.

26. Ibid.

27. Ibid.

28. Handwritten memo with Mar. 1947 FCC report (section titled "Whither") in fldr. "Color Television: Standards, Preliminary Releases," b. 2, Col. TV Hist./FCC, NA.

29. Chisholm, "CBS Color Venture," 300.

30. Ibid., 297–305.

31. Ibid., 330–33. On CBS color research during 1948 and 1949, see Frank Stanton to Robert Jones, Aug. 25, 1949, fldr. "Television," b. 6, TV Hist./FCC, NA.

32. Chisholm, "CBS Color Venture," 337–42. For statement by Johnson about wanting to make sure the commission did not ignore opportunities to prevent monopoly control, see letter quoting Johnson: Wayne Coy to Edwin Johnson, Feb. 1949, fldr. "Television Allocations, Pertinent Releases, 1946–49," b. 6, TV Hist./FCC, NA. For other Johnson quotations, see Edwin Johnson to David Sarnoff, May 13, 1949, b. 69, docket 8975, Dock/FCC, NA.

33. Chisholm, "CBS Color Venture," 369–73.

34. Wayne Coy to Edwin Johnson, Feb. 1949.

35. Frank Stanton to Robert Jones, Aug. 25, 1949.

36. For a description of RCA's dot-sequential system, see "First Report (Color Television Issues)," Sept. 1, 1950, 22–25.

37. Ibid., Sept. 1, 1950, 13–14. On CBS's field tests in homes, see Chisholm, "CBS Color Venture," 376.

38. Robert Jones testimony in FCC proceedings report, "Amendment of Section 3.606 of the Commission's Rules and Regulations . . . Utilization of Frequencies in the Band 470 to 890 Mcs. For Television Broadcasting," Sept. 26, 1949, b. 95, 2085, 2091, docket 8736 (also includes 8975, 9175, and 8976), Dock/FCC, NA.

39. For first Jones quote, see ibid., Sept. 27. For second quote of Robert Jones, see testimony ibid., May 4, 1950, b. 108, 10582. On membership of JTAC committee and views of Robert Jones, see Fink testimony ibid., Sept. 26, 1949, 2080.

40. For first Fink testimony see ibid., 2016, 2084, 2099–100. For last quote, see ibid., Apr. 5, 1950, b. 104, 7966.

41. Fink testimony ibid., Apr. 5, 1950, 7960.

42. Commissioner Jones quoted ibid., Mar. 16, 1950, b. 102, 6795.

43. Goldin testimony, ibid., 6793–98.

44. Goldin testimony, ibid., 6798–99, 6811, 6829–32.

45. Goldin testimony, ibid., 6820.

46. Jones quoted ibid., 6819–20. For view of company supporting hobbyists, see R. Robins to FCC, Nov. 4, 1949, fldr. "Television Hearings, 1928–45," b. 9, TV Hist./FCC, NA.

47. On industry resistance, see Chisholm, "CBS Color Venture," 353–55. Sarnoff quoted on page 355. Jones testimony in FCC, "Amendment of Section 3.606 of the Commission's Rules and Regulations . . . Utilization of Frequencies in the Band 470 to 890 Mcs. For Television Broadcasting," May 2, 1950, b. 108, 9822–23, docket 8736 (also includes 8975, 9175, and 8976), Dock/FCC, NA.

48. RMA testimony, ibid. Philco testimony ibid., Apr. 10, 1950, b. 105, 8173. CTI testimony ibid., May 17, 1950, b. 110, 11342.

49. Jones testimony ibid., May 4, 1950, 10291. CBS testimony ibid., Oct. 4, 1949, b. 96, 2970–71.

50. Engstrom testimony ibid., May 8, 1950, b. 108, 10914. Sarnoff testimony ibid., May 4, 1950, 10403.

51. On attempt to convince the commission to give the new NTSC the same authority as the old, see W. R. G. Baker to Wayne Coy, Jan. 24, 1950, fldr. "NTSC Correspondence", b. 5, TV Hist./FCC, NA. Baker testimony in FCC, "Amendment of Section 3.606 of the Commission's Rules and Regulations . . . Utilization of Frequencies in the Band 470 to 890 Mcs. For Television Broadcasting," May 1, 1950, b. 108, 9631–35, docket 8736 (also includes 8975, 9175, and 8976), Dock/FCC, NA.

52. Baker testimony ibid., May 1, 1950, 9633–36, 9666.

53. FCC counsel testimony ibid., May 1, 1950, 9664.

54. CBS, Baker, and FCC testimony ibid., 9733, 9739–40, 9742, 9773.

55. Jones testimony ibid., May 2, 1950, 9824.

56. "The Present Status of Color Television: Report of the Advisory Committee on Color Television to the Committee on Interstate and Foreign Commerce United States Senate," iv–v, 1, copy in fldr. "Color Television: Reports and Publications," b. 2, Col. TV Hist./FCC, NA.

57. Ibid., 1.

58. Ibid., 37, 40.

59. Ibid., 37, 54.

60. Fink testimony in FCC, "Amendment of Section 3.606 of the Commission's Rules and Regulations . . . Utilization of Frequencies in the Band 470 to 890 Mcs. For Television Broadcasting," Apr. 5, 1950, 7994. "Present Status of Color Television," 36–38.

61. "First Report (Color Television Issues)," Sept. 1, 1950, 57.

62. Ibid., 47.

63. Ibid., 48. Hecht testimony in FCC, "Petition of [CBS] for Changes in Rules Concerning Television Broadcast Stations," Feb. 10, 1947, vol. 8, b. 803, 1355, docket 7896, Dock/FCC, NA.

64. "First Report (Color Television Issues)," 49.

65. Ibid., 49–50.

66. Ibid., 50–52.

67. Ibid., 53—Jones's statement on page A-14.

68. Ibid., 53–55.

69. Ibid., 54–55.

70. Memo "Recommendations of the Office of Chief Engineer Regarding a Television Decision," June 23, 1950, 1, fldr. "Bracket Standards, 1950," b. 3, Col. TV Hist./FCC, NA.

71. App. C in memo "Initial Conclusions and Recommendations Concerning Color Television," to chief engineer from "Engineering Group Assigned to Consider the Color Television Problem," July 7, 1950, fldr. "Color Television: Conclusions and Recommendations Concerning Color Television, Chief Engineer, 1950," b. 1, Col. TV Hist./FCC, NA.

72. Memo "Engineering Department Policies Used for Processing Television Applications," n.d. (early 1950s), fldr. "Television History," b. 8, TV Hist./FCC, NA. Chapin's adapter is discussed in FCC, "Amendment of Section 3.606 of the Commission's Rules and Regulations . . . Utilization of Frequencies in the Band 470 to 890 Mcs. for Television Broadcasting," Nov. 22, 1949, b. 101, 5980. On important role of staff members, also see Erwin G. Krasnow, Lawrence D. Longley, and Herbert A. Terry, *The Politics of Broadcast Regulation* (New York: St Martin's Press, 1982), 35–37, 39.

73. "Organization and Management Survey of [FCC] for the Bureau of the Budget," vol. 1, 30–31. Copy in the Library of Congress. On FCC's rule-making process, including the different hearings conducted, see 32.

74. Untitled memo, chief of Technical Research Division to chief engineer, July 10, 1950, fldr. "TV Assignment and Planning: 1950–51 (2 of 2)," b. 8, TV Hist./FCC, NA.

75. Ibid.

76. On reaction of industry to bracket standards, see untitled memo, chief engineer to commissioners, n.d., fldr. "Color Decision: Bracket Standards, 1 of 2," b. 1, Col. TV Hist./FCC, NA. Chisholm, "CBS Color Venture," 357–59.

77. Chisholm, "CBS Color Venture," 401–5.

78. For records of RCA lawsuit, see fldr. "Color Television: Engineering Standards, Civil Action, 1950," b. 1, Col. TV Hist./FCC, NA. Chisholm, "CBS Color Venture," 406–10.

79. Chisholm, "CBS Color Venture," 413–15.

80. Ibid., 415–23, 428.

81. Ibid., 427.

82. On the color work of the NTSC during the early 1950s, see boxes 4 and 5, TV Hist./FCC, NA. Donald G. Fink, "Perspectives on Television: The Role Played by the Two NTSC's in Preparing Television Service for the American Public," *Proceedings of the IEEE* 64 (Sept. 1976): 1322–31.

Epilogue

1. On the dilemmas of corporate liberalism, see esp. Streeter, *Selling the Air* (see preface, n. 5).

2. For quotations, see Donald MacKenzie, *Inventing Accuracy: A Historical Sociology of Nuclear Missile Guidance* (Cambridge: MIT Press, 1990), 28, 411. Hughes, *Networks of Power* (see chap. 2, n. 42).

3. Christopher H. Sterling and John Michael Kittross, *Stay Tuned: A Concise*

History of American Broadcasting (Belmont, Calif.: Wadsworth, 1990), 381–84, 429–31, 468–70, 478.

4. Jonathan F. Galloway, *The Politics and Technology of Satellite Communications* (Lexington, Mass.: Lexington Books, 1972).

5. Richard W. Nelson, "Domestic Satellite Communication: Economic Issues in a Regulated Industry Undergoing Technical Change," in *Economic and Policy Problems in Satellite Communications*, ed. Joseph N. Pelton and Marcellus S. Snow (New York: Praeger, 1977).

6. Inglis, *Behind the Tube* (see chap. 4, n. 2), 416–32.

7. On development of high-definition television, see Joel Brinkley, *Defining Vision: How Broadcasters Lured the Government into Inciting a Revolution in Television* (New York: Harcourt Brace, 1997).

8. Sterling and Kittross, *Stay Tuned*, 517, 527; Patricia Aufderheide, *Communications Policy and the Public Interest: The Telecommunications Act of 1996* (New York: Guilford Press, 1999), 37–44.

9. Sterling and Kittross, *Stay Tuned*, 528; Robert Britt Horwitz, *The Irony of Regulatory Reform: The Deregulation of American Telecommunications* (New York: Oxford Univ. Press, 1989), 5–6.

10. On the recognition of the hybrid nature of decision making, see Jasanoff, *Fifth Branch* (see preface, n. 4).

11. Sterling and Kittross, *Stay Tuned*, 433–35; Horwitz, *The Irony of Regulatory Reform*, 19, 265.

)))

NOTE ON SECONDARY SOURCES

The literature on broadcasting and communications in the twentieth century is extensive; however, most of these works lack a historical focus. For an overall survey of the history of broadcasting in the United States, see the three-volume work by Erik Barnouw, *A History of Broadcasting in the United States* (New York: Oxford Univ. Press, 1966–70). Two other important historical surveys by communications experts are Christopher Sterling and John Michael Kittross, *Stay Tuned: A Concise History of American Broadcasting* (Belmont, Calif.: Wadsworth, 1990) and Sydney W. Head, Christopher H. Sterling, and Lemuel B. Schofield, *Broadcasting in America: A Survey of Electronic Media* (Boston: Houghton Mifflin, 1998). For a survey of technical developments, also see Andrew F. Inglis, *Behind the Tube: A History of Broadcasting Technology and Business* (Boston: Focal Press, 1990).

The quality of scholarship for the early history of radio is better than for other periods in the history of broadcasting. The best books include Susan J. Douglas, *Inventing American Broadcasting, 1899–1922* (Baltimore: Johns Hopkins Univ. Press, 1987); Robert W. McChesney, *Telecommunications, Mass Media, and Democracy: The Battle for the Control of U.S. Broadcasting, 1928–1935* (New York: Oxford Univ. Press, 1993); Susan Smulyan, *Selling Radio: The Commercialization of American Broadcasting, 1920–1934* (Washington, D.C.: Smithsonian Institution Press, 1994); and two works by Hugh G. J. Aitken, *Syntony and Spark: The Origins of Radio* (New York: Wiley, 1976) and *The Continuous Wave: Technology and American Radio, 1900–1932* (Princeton: Princeton Univ. Press, 1985). For an important book dealing with early radio policy, see Philip T. Rosen, *The Modern Stentors: Radio Broadcasting and the Federal Government, 1920–1934* (Westport, Conn.: Greenwood Press, 1980). For an international perspective, see

Daniel R. Headrick, *The Invisible Weapon: Telecommunications and International Politics, 1851–1945* (New York: Oxford Univ. Press, 1991). A good popular history of radio is Tom Lewis, *Empire of the Air: The Men Who Made Radio* (New York: HarperCollins, 1991).

The two main books on FM radio generally do not provide a critical perspective: see Don V. Erickson, *Armstrong's Fight for FM Broadcasting: One Man vs. Big Business and Bureaucracy* (University: Univ. of Alabama Press, 1973) and Lawrence Lessing, *Man of High Fidelity: Edwin Howard Armstrong* (Philadelphia: Lippincott, 1956). In this regard, see also the articles in "The Legacies of Edwin Howard Armstrong," *Proceedings of the Radio Club of America* 64 (1990), especially David L. Morton's essay, which calls for a more sophisticated, contextual analysis of the history of Armstrong and FM. One of the few examples of an author strikingly critical of the standard history, which generally portrays Armstrong in heroic terms and RCA as the evil corporation, is Inglis, *Behind the Tube*, 113–54. Inglis is a retired engineer who was employed by RCA for thirty years.

The best book on the early history of television in the United States is Joseph H. Udelson, *The Great Television Race: A History of the American Television Industry, 1925–1941* (University: Univ. of Alabama Press, 1982), 136–37. Udelson's work provides a useful overview of the television industry, but it does not focus on the regulation of television; nor does it analyze, using archival sources, questions relating to the role of technical experts and technical expertise in decision making. On the role of the Federal Communications Commission, see Robert H. Stern, *The Federal Communications Commission and Television: The Regulatory Process in an Environment of Rapid Technical Innovation* (New York: Arno Press, 1979), a reproduction of the author's 1950 dissertation. Another useful dissertation is John Michael Kittross, "Television Frequency Allocation Policy in the United States" (Ph.D. diss., University of Illinois, 1959; printed by Arno Press in 1979 under the same title). Both works are mainly based on secondary sources.

For general histories of television, see Erik Barnouw, *Tube of Plenty: The Development of American Television* (New York: Oxford Univ. Press, 1975); David E. Fisher and Marshall Jon Fisher, *Tube: The Invention of Television* (San Diego, Calif.: Harcourt Brace, 1997); Albert Abramson, *The History of Television, 1880 to 1941* (Jefferson, N.C.: McFarland, 1987); George Everson, *The Story of Television: The Life of Philo T. Farnsworth* (New York: Arno Press, 1974); and Albert Abramson, *Zworykin, Pioneer of Tele-*

vision (Chicago: Univ. of Illinois Press, 1995). For a technical history of television, see R. W. Burns, *Television: An International History of the Formative Years* (London: IEE Press, 1998). On special aspects of the history of television and politics, see James L. Baughman, *Television's Guardians: The FCC and the Politics of Programming, 1958–1967* (Knoxville: Univ. of Tennessee Press, 1985) and Erwin G. Krasnow, Lawrence D. Longley, and Herbert A. Terry, *The Politics of Broadcast Regulation* (New York: St. Martin's Press, 1982). For a history of color television in the United States, see Bradley Francis Chisholm, "The CBS Color Television Venture: A Study of Failed Innovation in the Broadcasting Industry," a 1987 Ph.D. dissertation at the University of Wisconsin–Madison. On public broadcasting, see George Gibson, *Public Broadcasting: The Role of the Federal Government, 1912–1976* (New York: Praeger Press, 1977) and Ralph Engelman, *Public Radio and Television in America: A Political History* (London: Sage, 1996).

This book builds on recent scholarship in the history of technology that engages general historical analysis and explores the multifaceted dynamics and detailed contingencies at the intersection of technology and society. Instead of writing master narratives that assume simple relationships involving the rational workings of a market economy or the independent determinative role of technological innovations on society, these studies focus on such complex issues as the multifaceted decisions of individual actors representing distinct institutions that helped shape technology. See the historiographic discussion in Merritt Roe Smith and Leo Marx, eds., *Does Technology Drive History? The Dilemma of Technological Determinism* (Cambridge: MIT Press, 1994), especially ix–xv, and articles by M. R. Smith and Philip Scranton in the same volume. On the importance of viewing technological development as simultaneously social, economic, and political, see especially Thomas P. Hughes, "The Seamless Web: Technology, Science, Etcetera, Etcetera," *Social Studies of Science* 16 (1982): 281–92. Also see introductory comments and important essays in Wiebe E. Bijker and John Law, eds., *Shaping Technology/Building Society: Studies in Sociotechnical Change* (Cambridge: MIT Press, 1992) and in Wiebe E. Bijker, Thomas P. Hughes, and Trevor J. Pinch, eds., *The Social Construction of Technological Systems: New Directions in the Sociology and History of Technology* (Cambridge: MIT Press, 1987).

Although in many cases it is important to analyze the differences between science and technology or science and engineering, my book focuses on the interrelationships. These have become especially significant in the twentieth century. I do not assume that technology is simply applied science. For a recent historiographic discussion of the relationship between science and technology that also emphasizes an "interactive model," see Edwin T. Layton Jr., "Through the Looking Glass; or, News from Lake Mirror Image," in *In Context: History and the History of Technology; Essays in Honor of Melvin Kranzberg,* ed. Stephen H. Cutcliffe and Robert C. Post, 29–52 (Bethlehem, Pa.: Lehigh Univ. Press, 1989).

The analysis in this book is informed by studies that emphasize that boundaries between such dichotomies as *technology* and *politics* need to be understood as contested and negotiated. On the importance of rhetorical strategies and what is called "boundary work" in science and technology studies, see especially Thomas F. Gieryn, "Boundary Work and the Demarcation of Science from Non-Science: Strains and Interests in Professional Ideologies of Scientists," *American Sociological Review* 48 (1983): 781–95; Sheila S. Jasanoff, "Contested Boundaries in Policy-Relevant Science," *Social Studies of Science* 17 (1987): 195–230; Hugh Richard Slotten, "The Dilemmas of Science in the United States: Alexander Dallas Bache and the U.S. Coast Survey," *Isis* 84 (1993): 26–49; Ronald Kline, "Constructing 'Technology' as 'Applied Science': Public Rhetoric of Scientists and Engineers in the United States, 1880–1945," *Isis* 86 (1995): 194–221. An important case study that explores boundary work and has influenced this book is Donald MacKenzie, *Inventing Accuracy: A Historical Sociology of Nuclear Missile Guidance* (Cambridge: MIT Press, 1990).

For general discussions of technology, science, and public policy, mainly in the United States, see especially Dorothy Nelkin, "Technology and Public Policy," in *Science, Technology, and Society: A Cross-disciplinary Perspective,* ed. Ina Spiegel-Rösing and Derek J. de Solla Price (Beverly Hills: Sage, 1977); Sheila Jasanoff, *The Fifth Branch: Science Advisers as Policymakers* (Cambridge: Harvard Univ. Press, 1990); Dorothy Nelkin, ed., *Controversy: Politics of Technical Decisions,* 3rd. ed. (London: Sage, 1992); Sheila Jasanoff, *Science at the Bar: Law, Science, and Technology in America* (Cambridge: Harvard Univ. Press, 1995). Also see Colleen A. Dunlavy, *Politics and Industrialization: Early Railroads in the United States and Prussia* (Princeton: Princeton Univ. Press, 1994).

Other studies of technology, science, and public policy particularly relevant to this work include Nicholas H. Steneck, *The Microwave Debate* (Cambridge: MIT Press, 1991); Brian Balogh, *Chain Reaction: Expert Debate and Public Participation in American Commercial Nuclear Power, 1945–1975*

(Cambridge: Cambridge Univ. Press, 1991); Susan Wright, *Molecular Politics: Developing American and British Regulatory Policy for Genetic Engineering, 1972–1982* (Chicago: Univ. of Chicago Press, 1994); S. P. Hays, *Beauty, Health, and Permanence: Environmental Politics in the United States, 1955–1985* (New York: Cambridge Univ. Press, 1987); Bruce Seely, *Building the American Highway System: Engineers as Policymakers* (Philadelphia: Temple Univ. Press, 1988); Pamela E. Mack, *Viewing the Earth: The Social Construction of the Landsat Satellite System* (Cambridge: MIT Press, 1990).

On the importance of corporate liberalism as a dominant political and economic ideology that originated during the early twentieth century, see especially R. Jeffrey Lustig, *Corporate Liberalism: The Origins of Modern American Political Theory, 1890–1920* (Berkeley: Univ. of California Press, 1990); Robert H. Wiebe, *The Search for Order, 1877–1920* (New York: Hill & Wang, 1967); Martin J. Sklar, *The Corporate Reconstruction of American Capitalism, 1890–1916: The Market, the Law, and Politics* (New York: Cambridge Univ. Press, 1988). On the importance of corporate liberalism to the history of broadcasting in the United States, see Thomas Streeter, *Selling the Air: A Critique of the Policy of Commercial Broadcasting in the United States* (Chicago: Univ. of Chicago Press, 1996). Ronald Kline's study of the electrical engineer Charles Steinmetz is an especially important study of the relationship between corporate liberalism and engineering; see Ronald R. Kline, *Steinmetz: Engineer and Socialist* (Baltimore: Johns Hopkins Univ. Press, 1992).

On the close relationship between private and public sectors, see Brian Balogh, "Reorganizing the Organizational Synthesis: Federal-Professional Relations in Modern America," *Studies in American Political Development* 5 (spring 1991): 119–72. Influential books by authors in the growing field of policy history include Thomas K. McCraw, *Prophets of Regulation: Charles Francis Adams, Louis D. Brandeis, James M. Landis* (Cambridge: Harvard Univ. Press, 1984); Ellis Wayne Hawley, *The New Deal and the Problem of Monopoly: A Study in Economic Ambivalence* (New York: Fordham Univ. Press, 1995); Morton Keller, *Regulating a New Society: Public Policy and Social Change in America, 1900–1933* (Cambridge: Harvard Univ. Press, 1994); Thomas K. McCraw, Morton Keller, et al., eds. *Regulation in Perspective: Historical Essays* (Cambridge: Harvard Univ. Press, 1981).

On the importance of technocratic values, see especially essays in Smith and Marx, eds., *Does Technology Drive History?*; John G. Gunnell, "The Technocratic Image and the Theory of Technocracy," *Technology and Culture* 23 (July 1982): 392–417; John M. Jordan, *Machine-Age Ideology:*

Social Engineering and American Liberalism, 1911–1939 (Chapel Hill: Univ. of North Carolina Press, 1994). On related issues of technocracy and scientific management, see Hugh G. J. Aitken, *Scientific Management in Action: Taylorism at Watertown Arsenal, 1908–1915* (Princeton: Princeton Univ. Press, 1985); Samuel Haber, *Efficiency and Uplift: Scientific Management in the Progressive Era, 1890–1920* (Chicago: Univ. of Chicago Press, 1964); William E. Akin, *Technocracy and the American Dream: The Technocracy Movement, 1900–1941* (Berkeley: Univ. of California Press, 1977).

On the cultural authority of science and technology see Marcel C. LaFollette, *Making Science Our Own: Public Images of Science, 1910–1955* (Chicago: Univ. of Chicago Press, 1990); Hugh Richard Slotten, "Humane Chemistry or Scientific Barbarism? American Responses to World War I Poison Gas, 1915–1930," *Journal of American History* 77 (1990): 476–98; Merritt Roe Smith, "Technology, Industrialization, and the Idea of Progress in America," in *Responsible Science: The Impact of Technology on Society*, ed. Kevin B. Byrne (San Francisco: Harper & Row, 1986).

On engineering professionalism and ethics, see Edwin T. Layton, *The Revolt of the Engineers: Society Responsibility and the American Engineering Profession* (Baltimore: Johns Hopkins Univ. Press, 1986); Bruce Sinclair, *A Centennial History of the American Society of Mechanical Engineers, 1880–1980* (Toronto: Univ. of Toronto Press, 1980); Monte Calvert, *The Mechanical Engineer in America, 1830–1910: Professional Cultures in Conflict* (Baltimore: Johns Hopkins Univ. Press, 1967); A. Michal McMahon, *The Making of a Profession: A Century of Electrical Engineering in America* (New York: IEEE, 1984); Terry S. Reynolds, *Seventy-five Years of Progress: A History of the American Institute of Chemical Engineers, 1908–1983* (New York: The Institute, 1983); William Kornhauser, *Scientists in Industry: Conflict and Accommodation* (Berkeley: Univ. of California Press, 1962); Peter Meiksins, "The 'Revolt of the Engineers' Reconsidered," *Technology and Culture* 29 (1988): 219–46; Terry S. Reynolds, ed., *The Engineer in America: A Historical Anthology from Technology and Culture* (Chicago: Univ. of Chicago Press, 1991). For a critical view of the role of engineers in corporate capitalism, see David F. Noble, *America by Design: Science, Technology, and the Rise of Corporate Capitalism* (New York: Knopf, 1977).

For obvious reasons, there are few historical studies of policy issues connected to recent developments in broadcasting and telecommunications. For an excellent theoretical discussion of deregulation and its historical background, see Robert B. Horwitz, *The Irony of Regulatory Reform: The Deregulation of American Telecommunications* (New York: Oxford Univ.

Press, 1989). Two books analyzing the predicted convergence of broadcasting with telephone and computer technology provide helpful overviews from different points of view: Bruce M. Owen, *The Internet Challenge to Television* (Cambridge: Harvard Univ. Press, 1999) and Patricia Aufderheide, *Communications Policy and the Public Interest: The Telecommunications Act of 1996* (New York: Guilford Press, 1999).

)))

INDEX

Library of Congress Cataloging-in-Publication Data

Slotten, Hugh Richard.
Radio and television regulation : broadcast technology in the United States, 1920–1960 / Hugh R. Slotten.
p. cm.
Includes bibliographical references and index.
ISBN 0-8018-6450-X (acid-free paper)
1. Broadcasting policy—United States—History—20th century. 2. Broadcasting—United States—History—20th century. I. Title.
HE8689.8 .S56 2000
384.54'0973'0904—dc21 00-008153